Worked Examples
Revised
Higher
Mathematics

by

M. Kyle

ISBN 0 7169 3148 6

ROBERT GIBSON · Publisher
17 Fitzroy Place, Glasgow, G3 7SF.

INTRODUCTION

This book is designed for pupils to use independently as revision exercises for the Revised Higher Mathematics examination. The six papers enclosed contain questions similar to the short response items of Paper I. The papers are on "Knowledge and Understanding" and include questions on graphicacy of trigonometry and derivatives, completing the square and recurrence relations.

Much of the content is relevant to Scotvec Modules M2/A1, M2/A2 and M2/C1 and should prove helpful to students studying these topics.

Printed by Bell and Bain Ltd., Glasgow

ANALYSIS OF CONTENTS

	Test A	Test B	Test C	Test D	Test E	Test F
Geometry Straight line	13	4,10,19	7,15	3,12,15	14,16	2,12,15,20
Circle	3,15	3	4,16	4,17	3	10
Vectors	11,14	7,20	14,20	10,23	13	
'O' Grade — Credit			22			
Algebra * Completing/square	24	25			22	4
Functions $f(g(x))$	22	24			11	6
*Sketching		15		16	12	
Quadratics	24	25		21	25	24
Polynomials	12,19	6	8,13,23	14,20	1,10	18,19
† Recurrence relations	25	9,23	25	24	21	21
* Logs — exp	7,23	1,22	5,19,21	6,19,22	6,19	14,17
'O' Grade — Credit	17		18	1	4,18	5,11
Calculus Sketching $f'(x)$	4,18	5	11	9		1
Differentiation	8,16	2,21	10,17	7,18	9,17,23	7,22
Integration	9,20	12	1	13	20	8
Trig Diffn.		14	6		5	9
Trigonometry Max — min		18	12	8	8	13
Graph sketching	1	16	3	2		23
Addition formula	2		9	5	2	3
* K cos/K sin	21	8	24	25	24	25
* Angle of gradient	6	17				
'O' Grade — Credit	10	13			7	
Trig equations	5	11		11	15	16

* Change of emphasis in Revised Higher.

† New content.

MATHEMATICS (Revised)

INSTRUCTIONS TO CANDIDATES

READ CAREFULLY

1. Full credit will be given only where the solution contains appropriate working.
2. Calculators may be used.
3. Answers obtained by readings from scale drawings will not receive any credit.

FORMULAE LIST

The equation $x^2 + y^2 + 2gx + 2fy + c = 0$ represents a circle centre $(-g, -f)$ and radius $\sqrt{(g^2 + f^2 - c)}$.

The equation $(x - a)^2 + (y - b)^2 = r^2$ represents a circle centre (a, b) and radius r.

Scalar Product: $\mathbf{a.b} = |\mathbf{a}||\mathbf{b}| \cos \theta$ where θ is the angle between $\mathbf{a}$ and $\mathbf{b}$

OR

$\mathbf{a.b} = a_1b_1 + a_2b_2 + a_3b_3$ where $\mathbf{a} = \begin{pmatrix} a_1 \\ a_2 \\ a_3 \end{pmatrix}$ and $\mathbf{b} = \begin{pmatrix} b_1 \\ b_2 \\ b_3 \end{pmatrix}$

Trigonometric formulae:
$$\sin (A \pm B) = \sin A \cos B \pm \cos A \sin B$$
$$\cos (A \pm B) = \cos A \cos B \mp \sin A \sin B$$
$$\cos 2A = \cos^2 A - \sin^2 A$$
$$= 2 \cos^2 A - 1$$
$$= 1 - 2 \sin^2 A$$
$$\sin 2A = 2 \sin A \cos A$$

Table of standard derivatives:

$f(x)$	$f'(x)$
$\sin ax$	$a \cos ax$
$\cos ax$	$-a \sin ax$

Table of standard integrals:

$f(x)$	$\int f(x)dx$
$\sin ax$	$\dfrac{-1}{a} \cos ax + C$
$\cos ax$	$\dfrac{1}{a} \sin ax + C$

TEST PAPER A

1. In how many places does the graph of $f : x \to \cos 3x$ cross the x-axis.
$0 \leqslant x < 360$

Draw a rough sketch to illustrate your answer.

2. In a right angled triangle $\tan A = \dfrac{5}{3}$,

find

(a) the exact value of $\cos 2A$

(b) and show that $\cos 2A + \sin 2A = \dfrac{7}{17}$

3. A circle with equation $x^2 + y^2 - 8x + 11 = 0$ touches another circle at the point $(6, 1)$. Find the equation of this second circle if its radius is twice as long.

4. In the graph shown, for what values of x are the statements

$f(x) > 0$ and $f'(x) < 0$ both true.

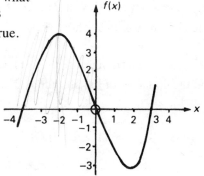

5. (a) Factorise fully $\quad 3\sin^2 x - \sin x - 2$

(b) Solve $3\sin^2 x - \sin x - 2 = 0$, for $0 \leqslant x \leqslant 360$.

6. Find the equation of the line through $(-1, 3)$ which makes an angle of $135°$ with the x-axis.

7. Solve for n, $2^n > 1500$ (answer to 2 decimal places).

8. When $f(x) = (x^2 - 3x)^3$. Find $f'(x)$ and $f'(-1)$.

9. If $\displaystyle\int_a^3 (3x^2 - 2x)dx = 20$ find a.

10. A chord AB is 3 units from the centre of a circle centre O and radius 5.

Find sin AÔB.

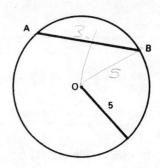

11. *(a)* A is the point $(3, 1, 4)$, B is the point $(6, 7, 10)$. P divides AB in the ratio 1:2 find the coordinates of P.

(b) State the ratio of AP:BP.

12. Find all the roots of the equation:

$f(x) = (x^2 + 1)(x^2 - 1)(x^2 + 3)(x^2 - 3), x \in R.$

State your answer in a solution set.

13. The vertices of a triangle are P$(2, -1)$, Q$(3, 2)$ and R$(6, 5)$.

Find the equation of the altitude AQ and the length of AQ (to 2 decimal places).

14. P $= (2, a, -3)$, Q $= (1, a, a)$. If OP is perpendicular to OQ, find the value of a.

15. *(a)* Find the coordinates of the centre and the length of the radius of the circle with equation

$x^2 + y^2 - 4x + 2y + 1 = 0.$

(b) Find the equation of this circle reflected in the x-axis.

16. Stationary values of the function $x^3 + mx$ occur when $x = \pm 1$ find the value of m.

17. Find the value of $(\sqrt{3} + 2\sqrt{2})^2$

18. The graph shown is $f(x)$.
Make a rough sketch of $f'(x)$.

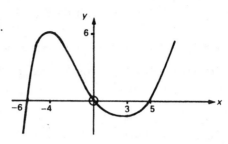

19. If the function $3x^3 - 16x^2 + px + 10$ is divisible by $(x - 1)$, find p and fully factorise the function.

20. If $f'(x) = 3x - 2$ and $f(2) = 7$, find $f(x)$.

21. Express $2 \sin x - 3 \cos x$ in the form $k \cos(x - \theta)$ and find the maximum and minimum values and the x, y, intercepts of the function for $0 \leqslant x \leqslant 360$.
Hence draw the graph of the function.

22. If $f(x) = 2x^2$ and $g(x) = 3x - 1$. Find $f(g(x))$.

23. *(a)* If £800 is invested at 7·2% per annum find the amount after 5 years.

(b) Find the difference in the amount if interest is calculated half yearly during this period.

24. *(a)* Show that the function $3x^2 - 4x + 2$, has no real roots.

(b) Show by completing the square that the function $3x^2 - 4x + 2$ has minimum value $\frac{2}{3}$.

(c) Make a rough sketch of the function.

25. *(a)* If $U_{r+1} = mU_r + c$ and $U_0 = 1$, $U_1 = -3$ and $U_2 = 21$.
Find m and c and state the relationship in the form $U_{r+1} = mU_r + c$.

(b) Find U_3 and U_{-1}.

(c) Find a value for U_r such that $U_{r+1} = U_r$.

TEST PAPER B

1. (a) If £300 is invested at 4·5% per annum find the amount after 5 years.

   (b) How many years would it take to double the investment at the same rate of interest?

2. Stationary values of the function $2x^3 + mx$ occur when $x = \pm 2$. Find the value of m. Hence state $f(x)$ and $f(-1)$.

3. (a) Find the coordinates of the centre and the length of the radius of the circle with equation
   $$x^2 + y^2 - 2x + 6y + 1 = 0.$$

   (b) State the equation of the circle after reflection in the y-axis.

4. The vertices of a triangle are A$(-1, 3)$, B$(2, -1)$ and C$(5, 4)$.
   Find the equation of the altitude BQ.

5. The graph of $f(x)$ is shown.
   Make a rough sketch of $f'(x)$.

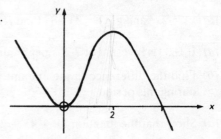

6. Find all the roots of the equation:
   $$f(x) = x(x^2 - 4)(x^2 - 2)(x^2 + 3), x \in R.$$
   State answer in a solution set.

7. (a) A is the point $(2, -1, 3)$, B is the point $(1, 6, -4)$.
   P divides AB in the ratio 2:-3 find the coordinates of P.

   (b) State the ratios AB:PB and AB:BP.

8. (a) Express $\cos x + \sin x$ in the form $k \cos (x - \alpha)$.

   (b) Solve the equation $\cos x + \sin x = 1, 0 \leqslant x \leqslant 2\pi$.

9. Janitors of a large school manage to clear 75% of the litter in the school grounds after school closes each day. Approximately 1 kg of litter is dropped daily. If there is 2 kg of litter in the grounds just before cleaning on a Monday how much litter will there be just after cleaning on Friday of the same week?

If this trend continues what is the eventual mass of the litter?

10. If the points $(1, 2)$, $(a, 4)$, $(b, 1)$ are collinear show that $a + 2b = 3$.

11. *(a)* Factorise fully $2 \cos^2 x + \cos x - 1$.

    (b) Solve $2 \cos^2 x + \cos x - 1 = 0$, for $0 \leqslant x \leqslant 360$.

12. If $\int_b^2 (3x^2 - 2)dx = 3$, find b.

13. In a right angled triangle $\tan A = \dfrac{3}{2}$ show that $\cos A$ can be expressed in the form $p\sqrt{13}$ and state the exact value of p.

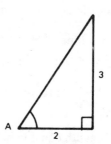

14. When $f(x) = \sin^3 x + \cos^2 x$. Find $f'(x)$ and $f'\left(\dfrac{\pi}{4}\right)$.

(Leave answer in surd form.)

15. The graph shown is $f(x)$. Sketch $-f(x)$ and $f(x) + 1$ on separate graphs.

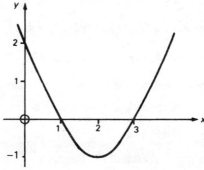

16. *(a)* In how many places does the graph of $f : x \rightarrow \sin 4x$ cross the x-axis. $0 \leqslant x \leqslant 360$.

(b) Sketch the graph to illustrate your answer.

17. *(a)* Find the equation of the line through $(0, -1)$ which makes an angle of $45°$ with the x-axis.

(b) State the equation of the line which is perpendicular to this line and passes through the point $(1, 3)$.

(c) State the y intercept of this perpendicular line and the angle which it makes with the x-axis.

18. *(a)* $A = 3 \cos \left(x - \frac{\pi}{6} \right)$ $\quad 0 \leqslant x \leqslant 2\pi$

Find the maximum and minimum values of A.

(b) State the coordinates of the turning points.

19. A triangle has coordinates $(1, 2)$, $(-3, 4)$ and $(5, 6)$ respectively.

Find the coordinates of the centroid of the triangle.

20. P is the point $(2, -1, 3)$ relative to rectangular axes OX, OY, OZ. Find the size of angle POX.

21. A cuboid with breadth equal to half its length has volume 576 cm^3.

Find the dimensions of the cuboid if the surface area is to be a minimum.

22. If $1 \cdot 2^x = 12$, find x to 2 decimal places.

23. *(a)* If $U_{r+1} = mU_r + c$. Find m and c and state the relationship in the form $U_{r+1} = mU_r + c$. When $U_0 = -1$, $U_1 = 7$, $U_2 = -9$.

(b) Find U_3 and U_{-1}.

(c) Find a value for U_r such that $U_{r+1} = U_r$.

24. If $g(x) = 3 - x^2$ and $f(x) = 1 - 2x$. Find $g(f(x))$.

25. *(a)* Show that the function $3x^2 - 2x + 5$, has no real roots.

(b) Show by completing the square that the function $3x^2 - 2x + 5$ has minimum value $\frac{14}{3}$.

(c) Make a rough sketch of the function.

TEST PAPER C

1. If $\int_a^6 (2x-3)dx = 20$ find a.

2. Write down an expression for the total shaded area as the sum of two integrals.

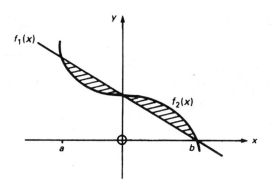

3. In how many places does the graph of $f:x \to \sin 3x$ cross the x-axis. $0 \leqslant x < 360$.

   Draw a rough sketch to illustrate your answer.

4. A circle has equation $x^2 + y^2 - 8x + 6y + 21 = 0$. Find the equation of the circle under reflection in the y-axis.

5. Solve for n, $0\cdot3^n < 0\cdot02$.

6. Find $f'\left(\frac{\pi}{4}\right)$ if $f(x) = 2 \sin 3x$.

7. A triangle has coordinates P(1, 2), Q(6, 3) and R(5, −2) respectively.

   (a) Find the coordinates of the centroid of the triangle.

   (b) Find the mid points M of PR and N of QP and the ratios QC:QM and RC:CN.

8. Factorise $x^3 - 3x^2 - 4x + 12$

9. In a right angled triangle $\tan A = \frac{3}{7}$, find the exact value of $\cos 2A$.

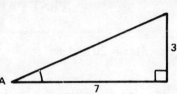

10. When $f(x) = (2x^2 - x)^5$. Find $f'(x)$.

11. $f(x)$ is shown in the diagram. Make a rough sketch of $f'(x)$.

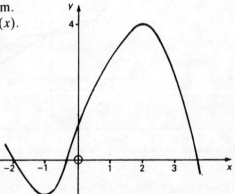

12. $C = 3 \cos \left(x + \frac{\pi}{3} \right)$ $0 \leqslant x \leqslant 2\pi$

Find the maximum and minimum values of C and the corresponding values of x.

13. Find all the roots of the equation:
$$f(x) = x(x + 2)(x^2 - 5)(x^2 + 1)(x^2 - 9), x \in R.$$
Give the answer in a solution set.

14. *(a)* A is the point $(1, -2, 4)$, B is the point $(-2, 4, 1)$. P divides AB in the ratio 2:1 find the coordinates of P.

(b) State the ratio AP:BP.

15. The vertices of a triangle are L(2, 4), M(−1, −2) and N(3, 7). Find the equation of the altitude LQ.

16. *(a)* Find the coordinates of the centre and the length of the radius of the circle with equation

$x^2 + y^2 - 6x + 8y + 9 = 0.$

(b) Find the equation of this circle after reflection in the x-axis.

17. Stationary values of the function $4x^3 + mx$ occur when $x = \pm \dfrac{\sqrt{3}}{2}$

(a) Find the value of m.

(b) State $f(x)$ and find $f(-2)$.

18. Find the value of $\quad (2\sqrt{3} - 5\sqrt{2})^2$

19. *(a)* If £500 is invested at 6·5% per annum find the amount after 10 years.

(b) Find the difference in the final amount if interest is added half yearly.

20. P is the point $(-2, 1, 3)$ relative to rectangular axes OX, OY, OZ. Find the size of angle $P\hat{O}Y$.

21. Find the least positive integer n for which $5^n > 9^{10}$

22. Find the locus of the point on the y-axis equidistant from the points $(7, 2)$ and $(-1, 3)$.

23. $f(x) = x^3 - 5x^2 - x + d$. If $f(x)$ is divisible by $(x + 1)$ find d and fully factorise the function.

24. Express $2\sin x - 5\cos x$ in the form $k\cos(x - \theta)$ and find the maximum and minimum values and the x, y, intercepts of the function for $0 \leqslant x \leqslant 360$.

Hence draw the graph of the function.

25. *(a)* If $U_{r+1} = mU_r + c$ and $U_0 = 3$, $U_1 = 2$ and $U_2 = 4$.

Find m and c and state the relationship in the form $U_{r+1} = mU_r + c$.

(b) Find U_3 and U_{-1}.

(c) Find a value for U_r such that $U_{r+1} = U_r$.

TEST PAPER D

1. Find the value of $(5 + 2\sqrt{3})^2$

2. In how many places does the graph of $f:x \rightarrow \cos 4x$ cross the x-axis. $0 \leqslant x \leqslant 180$.

   Draw a rough sketch to illustrate your answer.

3. If the points A(2, 1), B(a, 5), C(b, –7) are collinear show that $2a + b = 6$.

4. A circle has equation $x^2 + y^2 - 8x + 6y + 21 = 0$. Find the equation of the circle under reflection in the line $y = -x$.

5. In a right angled triangle $\tan A = \dfrac{1}{2\sqrt{2}}$, find the exact value of cos 2A and sin 2A.

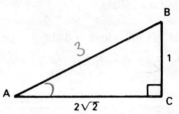

6. Solve for n, $0.5^n < 0.1$.

7. When $f(x) = (3x^2 - 2x)^4$. Find $f'(x)$ and $f'(-1)$.

8. $D = 2 \cos \left(x - \dfrac{\pi}{2}\right)$ $\qquad 0 \leqslant x \leqslant 2\pi$

   Find the maximum and minimum values of D and the values of x at these points.

9. The diagram shows the sketch of $f(x)$.

   Make a rough sketch of $f'(x)$.

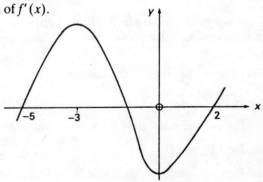

10. *(a)* A is the point (2, –2, 5), B is the point (–2, 2, –3). P divides AB in the ratio 1:3 find the coordinates of P.

   (b) State the ratio AB:PB.

11. *(a)* Factorise fully $3 \cos^2 x - 8 \cos x + 4$.

   (b) Solve $3 \cos^2 x - 8 \cos x + 4 = 0$, for $0 \leqslant x \leqslant 360$.

12. A triangle has coordinates (2, 5), (4, 1) and (8, –3) respectively. Find the coordinates of the centroid of the triangle.

13. If $\int_{a}^{2}(x^2 - 1)dx = 0$ find a.

14. Find all the roots of the equation:
$f(x) = x(x^2 - 3)(x^2 + 4)(x^2 - 1), x \in R$.
State answer in a solution set.

15. The vertices of a triangle are P(–2, 5), Q(2, –1) and R(4, 2). Find the equation of the altitude AQ.

16. Give two reasons why this graph cannot be the graph of $f(x) = 6 + x - x^2$.

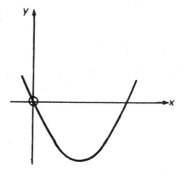

17. Find the coordinates of the centre and the length of the radius of the circle with equation:
$x^2 + y^2 + 12x - 4y + 15 = 0$.

18. Stationary values of the function $2x^3 + mx$ occur when $x = \pm\dfrac{1}{\sqrt{2}}$

   (a) Find the value of m.

   (b) State $f(x)$ and find $f(-3)$.

19. If £600 is invested at 6·2% per annum:

    (a) Find the amount after 4 years.

    (b) What rate would be required to double the amount of the investment over the same period?

20. If $x^3 - 3x^2 - x + a$ is divisible by $(x - 3)$, find a.

21. $f(x) = 3x^2 - 2x - c$. Find the value of c if the function has equal roots.

22. When y is plotted against x two points on the curve are $(1, 1·2)$ and $(5, 150)$. If the function is of the form $y = ax^n$. Find a and n and state the function.

23. $P = (1, 4, -1)$, $Q = (1, -2, 3)$. Find the cosine of angle POQ and state the size of the angle.

24. (a) If $U_{r+1} = mU_r + c$ and $U_0 = 2$, $U_1 = -1$ and $U_2 = 14$.

    Find m and c and state the relationship in the form $U_{r+1} = mU_r + c$.

    (b) Find U_3 and U_{-1}.

    (c) Find a value for U_r such that $U_{r+1} = U_r$.

25. Express $3 \sin x - 2 \cos x$ in the form $k \sin (x - \alpha)$ and find the maximum and minimum values and the x, y, intercepts of the function for $0 \leqslant x \leqslant 360$.

    Hence draw the graph of the function.

16

TEST PAPER E

1. If $x^3 + 4x^2 + x - t$ is divisible by $(x + 2)$, find t and fully factorise the function.

2. If $\tan A = K$. Prove that the exact value of $\cos 2A = \dfrac{1 - K^2}{1 + K^2}$

3. A circle has equation $x^2 + y^2 - 6x + 8y = 0$.

   (a) State the centre and radius of the circle.

   (b) Find the equation of the circle under reflection in the y-axis.

4. Solve for x: $\dfrac{2 + x}{2} - (2 - x) < 5$.

5. Find $f'\left(\dfrac{\pi}{6}\right)$ if $f(x) = 3 \sin 2x$.

6. Solve for x, $3^x = 100$. (Answer to 2 decimal places.)

7. In a right angled triangle $\tan A = \dfrac{1}{2}$, show that $\cos A$ can be expressed in the form $p\sqrt{5}$ and state the exact value of p

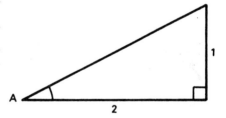

8. $H = 3 \cos\left(2x - \dfrac{\pi}{3}\right) \qquad 0 \leqslant x \leqslant 2\pi$

   Find the maximum and minimum values of H and the values of x where these occur.

9. When $f(x) = (2x + \sqrt{x})^3$. Find $f'(x)$ and $f'(4)$.

10. Find all the roots of the equation:

    $x(x + 2)(x^2 - 3)(x^2 + 1)(x^2 - 4)$, $x \in R$

    State the answer in a solution set.

11. If $h(x) = g(f(x))$ find $h(x)$ when $f(x) = 2x - 1$ and $g(x) = -x^2 + x + 2$.

12. The diagram shows the sketch of the function $f(x)$. Make a rough sketch of $-f(x)$ and $f(x) - 2$ (on two different sketches).

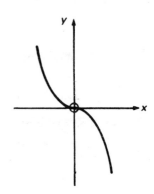

13. A is the point $(3, 1, 3)$, B is the point $(-2, 2, -2)$. P divides AB in the ratio 2:3 find the coordinates of P.

14. The vertices of a triangle are $P(2, 7)$, $Q(0, -1)$ and $R(-5, 4)$. Find the equation of the altitude AR.

15. *(a)* Factorise fully $6 \sin^2 x - \sin x - 2$.

(b) Solve $6 \sin^2 x - \sin x - 2 = 0$, for $0 \leqslant x \leqslant 360$.

16. If the points $(1, -1)$, $(a, 2)$, $(b, 1)$ are collinear show that $3b - 2a = 1$.

17. Stationary values of the function $4x^3 + mx$ occur when $x = \pm\frac{3}{2}$ find the value of m.

18. Find the value of $(2\sqrt{3} + 3\sqrt{2})^2$

19. If £1000 is invested at 5·8% per annum

(a) Find the amount after 7 years.

(b) Find the amount, if interest is added half yearly over this period.

20. Give an expression for the total sum of the shaded areas as the sum of two integrals.

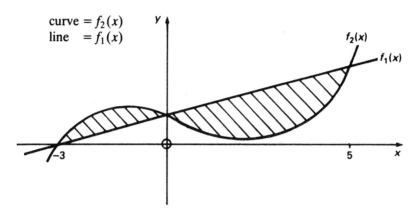

curve = $f_2(x)$
line = $f_1(x)$

21. A sequence is given by the recurrence relation $U_{r+1} = kU_r + t$.

(a) Find k and t if $U_0 = 0$, $U_1 = 2$ and $U_2 = -4$.

(b) Find U_4 and U_{-1}.

22. Show by completing the square that the function $4x^2 + 4x + 5$ has minimum value 4.

23. Find the equation of the tangent to the curve $y = 3x^2 - 2x + 1$, which is parallel to the line with equation $y = x - 3$.

24. Express $-4 \sin x - 3 \cos x$ in the form $k \cos (x - \lambda)$ and find the maximum and minimum values and the x, y, intercepts of the function for $0 \leqslant x \leqslant 360$.

Hence draw the graph of the function.

25. $f(x) = 2x^2 - bx + 3$. Find the value of b if the function

(a) has equal roots

(b) real roots

(c) no real roots

(d) make a sketch for each case.

19

TEST PAPER F

1. For what values of x is the function $x^3 - 3x - 5$ decreasing? Make a rough sketch to illustrate your answer.

2. If P and Q are points on the curve $3xy = -2$ with x coordinates 1 and -1 respectively, find the gradient of PQ.

3. By expressing $3x$ as $(2x + x)$ and x as $(2x - x)$. Find $\cos(3x) + \cos x$.

4. By the method of completing the square find the minimum value of $2x^2 + x + 2$.

5. If $\dfrac{x - 2y}{3} = \dfrac{y - 2x}{2}$. Find the value of $\dfrac{7x - 2y}{3x + y}$.

6. If $f(x) = x^2 - 3$ and $g(x) = 2 - x$ find $f(g(2))$.

7. The sketch shows the function $f(x)$. Make a rough sketch of $f'(x)$.

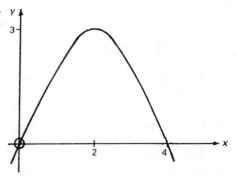

8. Evaluate $\displaystyle\int_1^4 \left(\sqrt{x} + \dfrac{1}{2\sqrt{x}} \right) dx$

9. Find $f'\left(\dfrac{\pi}{2}\right)$ if $f(x) = 2\cos 3x$.

10. A circle has equation $x^2 + y^2 - 6x + 8y = 0$.
    Find the equation of the circle under reflection in the x-axis.

11. If $\dfrac{x^2 + 100}{x + 10} = x - 10 + \dfrac{k}{x + 10}$. Find k.

12. If the points $(3, -1)$, $(a, 2)$, $(b, 5)$ are collinear show that $2a - b = 3$.

13. $K = 3 \cos \left(3x - \frac{\pi}{2} \right) \quad 0 \leqslant x \leqslant 2\pi$

Find the maximum and minimum values of D and the values of x where these occur.

14. *(a)* If £800 is invested at 7·8% per annum find the amount after 5 years.

(b) How long would it take to double the investment at the same rate of interest?

15. A triangle has coordinates $(-2, 5)$, $(2, -1)$ and $(6, 5)$ respectively. Find the coordinates of the centroid of the triangle.

16. Factorise fully $3 \sin^2 x - \sin x - 2$, and solve $3 \sin^2 x - \sin x - 2 = 0$, for $0 < x < 360$.

17. Solve for x, $e^x = 35$. (Answer to 2 decimal places.)

18. $f(x) = x(x^2 + 4)(x^2 - 3)(x^2 - 1)$, $x \in R$

Find the number of values of x for which $f(x) = 0$.

State these values in a solution set.

19. If $x^3 + 3x^2 - 4x + q$ is divisible by $(x - 2)$ find the value of q.

20. The vertices of a triangle are $P(-1, 5)$, $Q(-3, 2)$ and $R(9, -1)$. Find the equation of the altitude AP.

21. *(a)* $U_{r+1} = kU_r + t$, find k and t if $U_0 = 2$, $U_1 = -2$ and $U_2 = 10$.

(b) Find the value of U_r such that $U_{r+1} = U_r$.

22. A cuboid with length equal to three times the breadth has a volume of 1125 cm^3.

Find the dimensions of the cuboid if the surface area is to be minimised.

23. Sketch the graph of $f(x) = 3 \cos 2x$ for $0 \leqslant x \leqslant 2\pi$.

Show clearly the maximum and minimum values and the x, y intercepts.

21

24. $f(x) = ax^2 + 4x - 2$. Find the value of a if the function

    (a) has equal roots

    (b) real roots

    (c) no real roots

    (d) make a sketch for each case.

25. Express $3\cos x - \sin x$ in the form $k\cos(x - \alpha)$ and find the maximum and minimum values and the x, y, intercepts of the function for $0 \leqslant x \leqslant 360$.

Hence draw the graph of the function.

WORKED EXAMPLE — TEST PAPER A

1. $f(x) = \cos 3x$, $\{0 \leqslant x < 360\}$

Consider the pattern of $\cos x$ which repeats every 360°.

The period of $\cos x$ is 360°

The period of $\cos 3x$ is $\dfrac{360°}{3} = 120°$.

Hence the pattern repeats every 120°, almost 3 complete patterns occur within the given set $\{0 \leqslant x < 360\}$

x	0	30	60	90	120	150	180	210	240	270	300	330
$3x$	0	90	180	270	360	450	540	630	720	810	900	990
$\cos 3x$	1	0	−1	0	1	0	−1	0	1	0	−1	0

Points on the graph $(x, \cos 3x)$ are $(0°, 1)$ $(30°, 0)$ $(60°, -1)$ $(90°, 0)$ $(120°, 1)$ $(150°, 0)$ etc., as seen in the above table.

Hence the graph of $\cos 3x$ cuts the x-axis in 6 places namely $(30°, 0)$ $(90°, 0)$ $(150°, 0)$ $(210°, 0)$ $(270°, 0)$ $(330°, 0)$.

Graph of $\cos 3x$

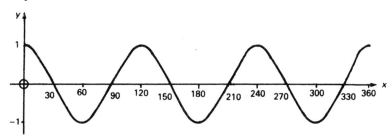

2. *(a)* $\text{Tan } A = \dfrac{5}{3}$ → sketch and label a right angled triangle.

Using the theorem of Pythagoras to find the 3rd side

$AB^2 = AC^2 + BC^2$

$\quad\quad = 5^2 + 3^2$

$AB^2 = 34$

$AB = \sqrt{34}$

Using the ratio of right angled triangles

$\cos A = \dfrac{3}{\sqrt{34}}$ $\quad$ $\sin A = \dfrac{5}{\sqrt{34}}$

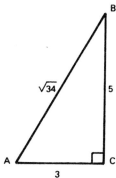

23

cos 2A   Using double angle Trig. formula

$$\cos 2A = \cos A \cos A - \sin A \sin A$$
$$= \cos^2 A - \sin^2 A$$

$\cos A = \dfrac{3}{\sqrt{34}}$, $\sin A = \dfrac{5}{\sqrt{34}}$, $\cos^2 A - \sin^2 A = \left(\dfrac{3}{\sqrt{34}}\right)^2 - \left(\dfrac{5}{\sqrt{34}}\right)^2$

$$= \dfrac{9}{34} - \dfrac{25}{34} = -\dfrac{16}{34}$$

$$= -\dfrac{8}{17}$$

OR using $\cos^2 A - \sin^2 A$
$$= \cos^2 A - (1 - \sin^2 A)$$
$$= 2\cos^2 A - 1$$

$\cos A = \dfrac{3}{\sqrt{34}}$, $= 2\left(\dfrac{3}{\sqrt{34}}\right)^2 - 1$

$$= 2\left(\dfrac{9}{34}\right) - 1 = \dfrac{9}{17} - \dfrac{17}{17}$$

$$\cos 2A = -\dfrac{8}{17}$$

(b) To show $\cos 2A + \sin 2A = \dfrac{7}{17}$

$\cos 2A = -\dfrac{8}{17}$ as found in part (a).

By double angle formula $\sin 2A = \sin (A + A)$
$$= \sin A \cos A + \cos A \sin A$$
$$= 2 \sin A \cos A$$

$\cos A = \dfrac{3}{\sqrt{34}}$, $\sin A = \dfrac{5}{\sqrt{34}}$,

$$\Rightarrow 2 \sin A \cos A = 2\left(\dfrac{5}{\sqrt{34}}\right)\left(\dfrac{3}{\sqrt{34}}\right)$$

$$= \dfrac{30}{34}$$

$\sin 2A = \dfrac{15}{17}$, and from part (a) $\cos 2A = -\dfrac{8}{17}$

hence $\cos 2A + \sin 2A = -\dfrac{8}{17} + \dfrac{15}{17}$

$$= \dfrac{7}{17} \text{ as given.}$$

3. Circle has general equation $x^2 + y^2 + 2gx + 2fy + c = 0$
with centre $(-g, -f)$ and radius $\sqrt{(g^2 + f^2 - c)}$

Given circle has equation $x^2 + y^2 - 8x + 11 = 0$

hence $2g = -8$ and $2f = 0$

so centre is $(4, 0)$

$c = 11$ radius is $\sqrt{(4^2 - 11)}$

$= \sqrt{5}$

centre $(4, 0)$ radius $\sqrt{5}$

The circle with radius twice as long has radius $2\sqrt{5}$

The distance between the two centres is $\sqrt{5} + 2\sqrt{5} = 3\sqrt{5}$

Let centre of small circle $= C_1$

Let centre of large circle $= C_2$

Let point of intersection $= P(6, 1)$

$\overrightarrow{C_1P}:\overrightarrow{PC_2}$ has ratio 1:2

$$\overrightarrow{C_1P} = p - c_1 = \begin{pmatrix} 6 \\ 1 \end{pmatrix} - \begin{pmatrix} 4 \\ 0 \end{pmatrix} = \begin{pmatrix} 2 \\ 1 \end{pmatrix}$$

$$\overrightarrow{PC_2} = 2\overrightarrow{C_1P} = 2\begin{pmatrix} 2 \\ 1 \end{pmatrix} = \begin{pmatrix} 4 \\ 2 \end{pmatrix}$$

$$\overrightarrow{OC_2} = \overrightarrow{OP} + \overrightarrow{PC_2} = \begin{pmatrix} 6 \\ 1 \end{pmatrix} + \begin{pmatrix} 4 \\ 2 \end{pmatrix} = \begin{pmatrix} 10 \\ 3 \end{pmatrix} \quad C_2 = (10, 3)$$

Hence centre of the larger circle $= (10, 3)$

radius $= 2\sqrt{5}$

Equation is $(x - 10)^2 + (y - 3)^2 = (2\sqrt{5})^2$

$\Leftrightarrow x^2 - 20x + 100 + y^2 - 6y + 9 = 20$

$\Leftrightarrow x^2 + y^2 - 20x - 6y + 89 = 0$

Alternative Method:

By using $-g = 10, -f = 3, c = 2\sqrt{5}$

in the general equation $x^2 + y^2 + 2gx + 2fy + c = 0$

$\Leftrightarrow x^2 + y^2 + 2(-10)x + 2(-3)y + c = 0$

$x^2 + y^2 - 20x - 6y + c = 0$

and now find c by solving $r = 2\sqrt{5}$

$r = \sqrt{g^2 + f^2 - c}$

$2\sqrt{5} = \sqrt{10^2 + 3^2 - c}$

$20 = 109 - c$

$\Rightarrow c = 89$

Hence equation of circle $= x^2 + y^2 - 20x - 6y + 89 = 0$

4.

$f(x) > 0, \; -4 < x < 0,$
$f'(x) < 0, -2 < x < 2$

$f(x)$ positive for
all points above
the x-axis

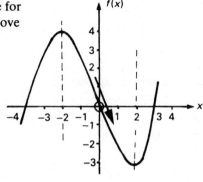

$f'(x) = m$, (gradient)
gradient negative
between $x = -2$ and
$x = 2$

$f(x) > 0$ and $f'(x) < 0$
$\Rightarrow \{x : -2 < x < 0, x \in R\}$

5. *(a)* $3 \sin^2 x - \sin x - 2 = 0,$ let $\sin x = x$ to simplify
$\Rightarrow 3 \sin^2 x - \sin x - 2 = 0$
$\Rightarrow 3x^2 - x - 2 = 0$
$\Rightarrow (3x + 2)(x - 1) = 0$

$3 \sin^2 x - \sin x - 2 = 0$
$\Rightarrow (3 \sin x + 2)(\sin x - 1) = 0$

(b) $\Rightarrow \sin x = -\dfrac{2}{3}$ or $\sin x = 1$

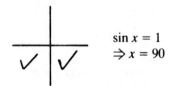

$\sin x = 1$
$\Rightarrow x = 90$

$\sin$ is $-$ve in quadrant 3 and 4

$\sin^{-1} \dfrac{2}{3} = 41 \cdot 8$

In ③ $= 180 + 41 \cdot 8 \Rightarrow x = 221 \cdot 8$
In ④ $= 360 - 41 \cdot 8 \Rightarrow x = 318 \cdot 2$

solution set $\{90°, 221 \cdot 8°, 318 \cdot 2°\}$

6.

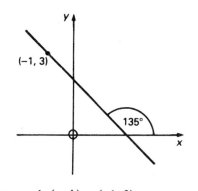

The gradient of the line which makes an angle of 135° with the x-axis is the tangent of 135° which is −1.

We can find the equation of any straight line when we know the gradient of the line and any point on the line by using the formula $y - b = m(x - a)$ where m is the gradient and (a, b) are co-ordinates of the point.

$m = -1, (a, b) = (-1, 3)$
$$y - b = m(x - a)$$
$$y - 3 = -1(x - (-1))$$
$$y - 3 = -1(x + 1)$$
$$y - 3 = -x - 1$$
$$y = -x + 2 \text{ is the required equation.}$$

7. $2^n > 1500$

* This can be solved by trial and error using a calculator.

Log method:
$$2^n = 1500$$
$$\log 2^n = \log 1500$$
$$\Rightarrow n \log 2 = \log 1500$$
$$\Rightarrow n = \frac{\log 1500}{\log 2}$$
$$\Rightarrow n = 10 \cdot 55 \text{ to 2 decimal places}$$
$$\text{check using } 2^{10 \cdot 55}, 2^{10 \cdot 54}, 2^{10 \cdot 56}$$
$$2^n > 1500 \Rightarrow n > 10 \cdot 55$$

8. $f(x) = (x^2 - 3x)^3$

Find $f'(x)$ by chain rule method.

$$f'(x) = 3(x^2 - 3x)^2(2x - 3)$$
$$= 3(2x - 3)(x^2 - 3x)^2$$
$$f'(-1) = 3(2(-1) - 3)((-1)^2 - 3(-1))^2$$
$$= 3(-5)(1 + 3)^2$$
$$= -15(4)^2$$
$$= -240$$
$$f'(-1) = -240$$

9. $\int_a^3 (3x^2 - 2x)\,dx = 20 \Rightarrow \left[\dfrac{3x^{2+1}}{2+1} - \dfrac{2x^{1+1}}{1+1}\right]_a^3 = 20$

$$= \left[\dfrac{3x^3}{3} - \dfrac{2x^2}{2}\right]_a^3 = 20$$

$$= [x^3 - x^2]_a^3 = 20$$

$$\Rightarrow (3^3 - 3^2) - (a^3 - a^2) = 20$$

$$\Rightarrow 18 - a^3 + a^2 = 20$$

$$\Rightarrow a^2 - a^3 = 2$$

By trial and error assume that $a < 3$,

$$\Rightarrow a^2(1 - a) = 2$$

$$2^2(1 - 2) \neq 2$$

$$(-1)^2(1 - (-1))$$

$$1(2) = 2$$

$$\Rightarrow a = -1$$

* By synthetic division using $-a^3 + a^2 = 2$ as $a^3 - a^2 + 2 = 0$

	a^3	a^2	a^1	a^0
-1	1	-1	0	2
		-1	$+2$	-2
	1	-2	$+2$	0

Since $a^3 - a^2 + 2$ is divisible by -1 then $(a + 1)$ is a factor
$$\Rightarrow a = -1$$

10.

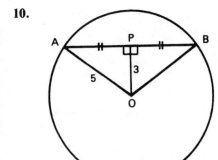

Let mid point of chord $AB = P$
Using the Theorem of Pythagoras on right angled triangle $P\hat{O}B$
$$OB = \text{radius} = 5, \;|OP| = 3$$
$$PB^2 = 5^2 - 3^2$$
$$PB^2 = 16 \Rightarrow PB = 4$$

Using ratio of right angled triangles $\cos P\hat{O}B = \dfrac{3}{5}$

$$\sin P\hat{O}B = \dfrac{4}{5}$$

To find sin AÔB

$$A\hat{O}B = 2\,P\hat{O}B$$

Let $A\hat{O}B = x$, hence $P\hat{O}B = 2x$

$$\sin 2x = 2\sin x \cos x$$

$$\sin x = \frac{4}{5}\,,\ \cos x = \frac{3}{5}$$

$$2\sin x \cos x$$

$$= 2\left(\frac{4}{5}\right)\left(\frac{3}{5}\right)$$

$$= 2\left(\frac{12}{25}\right)$$

$$= \frac{24}{25}$$

11. (a) A(3, 1, 4), B(6, 7, 10)

$$\underset{\sim}{a} = \begin{pmatrix} 3 \\ 1 \\ 4 \end{pmatrix},\ \underset{\sim}{b} = \begin{pmatrix} 6 \\ 7 \\ 10 \end{pmatrix}$$

$$\overrightarrow{AP}\!:\!\overrightarrow{PB}$$

$$1:2$$

$$\Rightarrow 2\overrightarrow{AP} = \overrightarrow{PB}$$

$$\overrightarrow{AP} = \underset{\sim}{p} - \underset{\sim}{a} \qquad\qquad \overrightarrow{PB} = \underset{\sim}{b} - \underset{\sim}{p}$$

$$2\overrightarrow{AP} = 2(\underset{\sim}{p} - \underset{\sim}{a}),\qquad \overrightarrow{PB} = (\underset{\sim}{b} - \underset{\sim}{p})$$

$$\Rightarrow 2\underset{\sim}{p} - 2\underset{\sim}{a} = \underset{\sim}{b} - \underset{\sim}{p}$$

$$3\underset{\sim}{p} = 2\underset{\sim}{a} + \underset{\sim}{b}$$

$$\Rightarrow 3\underset{\sim}{p} = 2\begin{pmatrix} 3 \\ 1 \\ 4 \end{pmatrix} + \begin{pmatrix} 6 \\ 7 \\ 10 \end{pmatrix}$$

$$\begin{pmatrix} 6 \\ 2 \\ 8 \end{pmatrix} + \begin{pmatrix} 6 \\ 7 \\ 10 \end{pmatrix} = \begin{pmatrix} 12 \\ 9 \\ 18 \end{pmatrix}$$

$$3\underset{\sim}{p} = \begin{pmatrix} 12 \\ 9 \\ 18 \end{pmatrix} \Rightarrow \underset{\sim}{p} = \begin{pmatrix} 4 \\ 3 \\ 6 \end{pmatrix}$$

$$\underset{\sim}{p} = \overrightarrow{OP} \qquad \text{Hence } P = (4, 3, 6)$$

Alternative Method:
This can also be solved using the section formula.

$$P = \frac{1}{3}(2a + b)$$

$$= \frac{1}{3}\left[\begin{pmatrix} 6 \\ 2 \\ 8 \end{pmatrix} + \begin{pmatrix} 6 \\ 7 \\ 10 \end{pmatrix}\right]$$

$$= \frac{1}{3}\begin{pmatrix} 12 \\ 9 \\ 18 \end{pmatrix} = (4, 3, 6)$$

(b) A(3, 1, 4), P(4, 3, 6), B(6, 7, 10)

$$\overrightarrow{AP} = \underset{\sim}{p} - \underset{\sim}{a} \qquad\qquad \overrightarrow{BP} = \underset{\sim}{p} - \underset{\sim}{b}$$

$$\underset{\sim}{p} - \underset{\sim}{a} = \begin{pmatrix} 4 \\ 3 \\ 6 \end{pmatrix} - \begin{pmatrix} 3 \\ 1 \\ 4 \end{pmatrix} \qquad \underset{\sim}{p} - \underset{\sim}{b} = \begin{pmatrix} 4 \\ 3 \\ 6 \end{pmatrix} - \begin{pmatrix} 6 \\ 7 \\ 10 \end{pmatrix}$$

$$= \begin{pmatrix} 1 \\ 2 \\ 2 \end{pmatrix} \qquad\qquad = \begin{pmatrix} -2 \\ -4 \\ -4 \end{pmatrix}$$

$$\overrightarrow{AP} : \overrightarrow{BP} = 1 : -2$$

12. $f(x) = (x^2 + 1)(x^2 - 1)(x^2 + 3)(x^2 - 3) \ x \in R$

Roots occur when $f(x) = 0$

$\Rightarrow \quad (x^2 + 1) = 0$ or $(x^2 - 1) = 0$, $(x^2 + 3) = 0$ or $(x^2 - 3) = 0$

$\quad x^2 + 1 = 0$

$\Rightarrow \quad\quad x^2 = -1$

No solution in real number system

$\quad x^2 - 1 = 0$

$\Rightarrow \quad\quad x^2 = 1$

$\Rightarrow \quad\quad x = \pm 1$

$\quad x^2 + 3 = 0$

$\Rightarrow \quad\quad x^2 = -3$

No solution in real numbers

$$x^2 - 3 = 0$$
$$\Rightarrow x^2 = 3$$
$$\Rightarrow x = \pm\sqrt{3}$$

Solution set $\{-\sqrt{3}, -1, 1, \sqrt{3}\}$

13. $P(2, -1)$, $Q(3, 2)$, $R(6, 5)$

gradient of PR $= \dfrac{5 - (-1)}{6 - 2}$

$$\frac{6}{4} = \frac{3}{2} = m$$

If line QA is perpendicular to line PR then $m_{QA} \cdot m_{PR} = -1$

$m_{PR} = \dfrac{3}{2}$ hence $m_{QA} = -\dfrac{2}{3}$

Point on QA = $Q(3, 2)$ gradient of QA $= -\dfrac{2}{3}$

$$y - b = m(x - a)$$
$$y - 2 = -\frac{2}{3}(x - 3)$$
$$3y - 6 = -2x + 6$$

$3y + 2x = 12$ is the equation of the altitude QA.

31

To find the length of QA need to find point A.
where QA intersects PR
Find equation of PR.

$m_{PR} = \dfrac{3}{2}$ shown, through point R(6, 5)

$$y - 5 = \frac{3}{2}(x - 6)$$
$$2y - 10 = 3x - 18$$
$$2y \quad\ = 3x - 8$$
$$\Rightarrow 2y - 3x = -8 \text{ the equation of PR}$$

Equation of QA: $3y + 2x = 12 \Rightarrow 9y + 6x = 36$
Equation of PR: $2y - 3x = -8 \Rightarrow 4y - 6x = -16$

$$13y = 20$$
$$y = \frac{20}{13}$$
$$3y + 2x = 12$$
$$y = \frac{20}{13}$$
$$\frac{60}{13} + 2x = 12$$
$$2x = 12 - \frac{60}{13}$$
$$x = \frac{1}{2}\left(12 - \frac{60}{13}\right)$$
$$x = 6 - \frac{30}{13}$$
$$x = \frac{48}{13} \text{ and } y = \frac{20}{13}$$

$A\left(\dfrac{48}{13}, \dfrac{20}{13}\right)$

To find length of QA

$Q(3, 2), A\left(\dfrac{48}{13}, \dfrac{20}{13}\right)$

$$|QA|^2 = \left(\frac{48}{13} - 3\right)^2 + \left(\frac{20}{13} - 2\right)^2$$
$$= \left(\frac{9}{13}\right)^2 + \left(\frac{-6}{13}\right)^2$$
$$\frac{81}{169} + \frac{36}{169} = \frac{117}{169}$$

$QA \qquad = \sqrt{\dfrac{117}{169}}$

$QA \qquad \approx 0{\cdot}83 \text{ (to 2 decimal places)}$

32

14. P(2, a, –3), Q(1, a, a)

$$\overrightarrow{OP} = \underset{\sim}{p} = \begin{pmatrix} 2 \\ a \\ -3 \end{pmatrix}, \overrightarrow{OQ} = \underset{\sim}{q} = \begin{pmatrix} 1 \\ a \\ a \end{pmatrix}$$

If $\overrightarrow{OP}$ is perpendicular to $\overrightarrow{OQ}$ then $p.q = 0$

$$\Rightarrow \begin{pmatrix} 2 \\ a \\ -3 \end{pmatrix} \cdot \begin{pmatrix} 1 \\ a \\ a \end{pmatrix} = 0$$

$\Rightarrow 2 + a^2 - 3a = 0$

$\Rightarrow a^2 - 3a + 2 = 0$

$\quad (a - 1)(a - 2) = 0$

$\quad a = 2 \text{ or } a = 1$

Check with $a = 2$; $\begin{pmatrix} 2 \\ 2 \\ -3 \end{pmatrix} \cdot \begin{pmatrix} 1 \\ 2 \\ 2 \end{pmatrix} = 2 + 4 - 6 = 0$

Check with $a = 1$; $\begin{pmatrix} 2 \\ 1 \\ -3 \end{pmatrix} \cdot \begin{pmatrix} 1 \\ 1 \\ 1 \end{pmatrix} = 2 + 1 - 3 = 0$

15. *(a)* $x^2 + y^2 - 4x + 2y + 1 = 0$

General equation of a circle $x^2 + y^2 + 2gx + 2fy + c = 0$

where centre $= (-g, -f)$

radius $\quad = \sqrt{(g^2 + f^2 - c)}$

$x^2 + y^2 - 4x + 2y + 1 = 0$

$2g = -4, 2f = 2$
$-g = 2, -f = -1, c = 1$

$\quad\quad$ radius $= \sqrt{(2^2 + 1^2 - 1)} = \sqrt{4} = 2$

Hence centre $= (2, -1)$, radius $= 2$

(b) reflected in the x-axis

centre = (2, 1)

radius unchanged

new equation is

$$x^2 + y^2 - 4x - 2y + 1 = 0$$

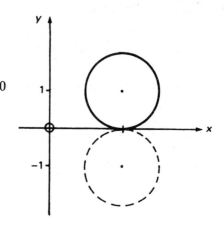

16. $f(x) = x^3 + mx$

$f'(x) = 3x^2 + m$

$3x^2 + m$ = gradient of tangent

gradient = 0 for stationary value

$$\Rightarrow 3x^2 + m = 0$$

$x = \pm 1 \Rightarrow 3(-1)^2 + m = 0$

$$3 + m = 0$$

$$\Rightarrow \qquad m = -3$$

$f(x) = x^3 - 3x$

17. $(\sqrt{3} + 2\sqrt{2})^2$

$= (\sqrt{3} + 2\sqrt{2})(\sqrt{3} + 2\sqrt{2})$

$= \sqrt{3}(\sqrt{3} + 2\sqrt{2}) + 2\sqrt{2}(\sqrt{3} + 2\sqrt{2})$

$= 3 + 2\sqrt{6} + 2\sqrt{6} + 8$

$= 11 + 4\sqrt{6}$

18.

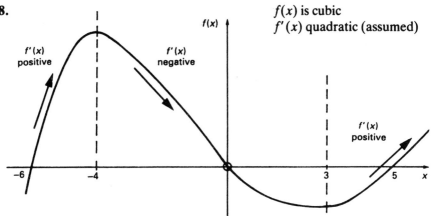

$f(x)$ is cubic
$f'(x)$ quadratic (assumed)

$f'(x)$ positive

$f'(x)$ negative

$f'(x)$ positive

	$x < -4$	$x = -4$	$-4 < x < 3$	$x = 3$	$x > 3$
$f'(x)$	+ve	0	−ve	0	+ve
Plot $f'(x)$ in relation to the x-axis	above	on	below	on	above

sketch of $f'(x)$

points
$(-4, 0)(3, 0)$

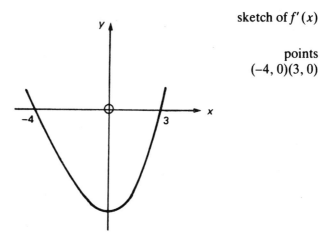

19. $f(x) = 3x^3 - 16x^2 + px + 10$

If $f(x)$ is divisible by $(x-1)$

then by synthetic division remainder $= 0$

$\div(x-1)$	x^3	x^2	x^1	x^0	
	3	-16	p	10	
1		3	-13	$-13 + p$	
	3	-13	$-13 + p$	$-3 + p$	(Remainder)

$$\Rightarrow \quad -3 + p = 0$$
$$p = 3$$

$3x^3 - 16x^2 + 3x + 10 \div (x-1)$
$= 3x^2 - 13x - 10$ (since $-13 + p = -13 + 3 = -10$)
factorised $= (3x + 2)(x - 5)$
fully factorised $f(x) = (x-1)(x-5)(3x+2)$

20. $f'(x) = 3x - 2$

$$f(x) = \frac{3x^2}{2} - 2x + c$$

$$f(2) = \frac{3(2)^2}{2} - 2(2) + c$$

$$= 6 - 4 + c$$

$$2 + c$$

given $f(2) = 7$
$\Rightarrow 2 + c = 7$
$\Rightarrow \quad c = 5$

Hence $f(x) = \frac{3x^2}{2} - 2x + 5$

21. $2\sin x - 3\cos x, \quad \{0 < x < 360\}$
$= -3\cos x + 2\sin x$

In general, $a\cos x + b\sin x$ has tangent $= \frac{b}{a}$ in quadrant where (a, b) is plotted.

$a = -3, b = 2$

$\Rightarrow \tan\theta = \frac{2}{-3} \quad (a, b) = (-3, 2)$

In quadrant ②

$\tan\theta^{-1}\frac{2}{3} = 33{\cdot}7$

$\theta = 180 - 33{\cdot}7$

$= 146{\cdot}3°$

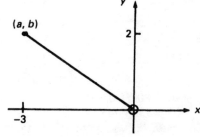

$k^2 = a^2 + b^2$
$k^2 = 2^2 + (-3)^2$
$\quad = 4 + 9 = 13$
$k = \sqrt{13}$
In form $k \cos(x - \theta)$
gives $\sqrt{13} \cos(x - 146 \cdot 3)$
max $= \sqrt{13}$ min $= -\sqrt{13}$
Period of $\cos x = 360°$
Graph is graph of $\cos x$
moved to right $146 \cdot 3°$
Points $(0, -3)$ $(56 \cdot 3, 0)$ $(146 \cdot 3, \sqrt{13})$
$(236 \cdot 3, 0)$ $(326 \cdot 3, -\sqrt{13})$ $(360, -3)$

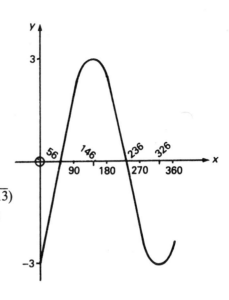

22. $\quad f(x) = 2x^2$
$\quad\quad g(x) = 3x - 1$
$\quad f(g(x)) = f(3x - 1)$
$\quad\quad\quad = 2(3x - 1)^2$
$\quad\quad\quad = 2(9x^2 - 6x + 1)$
$\quad f(g(x)) = 18x^2 - 12x + 2$

23. *(a)* $\quad$ £800 = Principal
$\quad\quad\quad n = 5$ (number of years)
$\quad\quad\quad r = $ rate $= 7 \cdot 2\%$
$\quad\quad\quad R = 1 + 7 \cdot 2\%$
$\quad\quad$ Amount $= PR^n$
$\quad\quad\quad\quad = 800(1 \cdot 072)^5$
$\quad\quad\quad\quad = $ £1132·57, (to nearest pence)

(b) Half yearly
Annual rate $7 \cdot 2\%$
Half yearly $3 \cdot 6\%$
Number of periods of 6 months over 5 years $= 10$
$A = PR^n$, $P = 800$, $R = 1 \cdot 036$, $n = 10$
$A = 800(1 \cdot 036)^{10}$
$\quad = $ £1139·43
Difference $= 1139 \cdot 43 - 1132 \cdot 57 = $ £6.86 more interest.

24. *(a)* $3x^2 - 4x + 2 = 0$

$a = 3, b = -4, c = 2$

$b^2 - 4ac < 0$ for no real roots

$(-4)^2 - 4(3)(2)$

$16 - 24 = -8$

since $\sqrt{-8} \in R$

function has no real roots.

(b) $\quad 3x^2 - 4x + 2$

$$3\left(x^2 - \frac{4}{3}x + \frac{2}{3}\right)$$

$$3\left[\left(x - \frac{2}{3}\right)^2 - \frac{4}{9} + \frac{2}{3}\right]$$

$$3\left[\left(x - \frac{2}{3}\right)^2 + \frac{2}{9}\right]$$

$$= 3\left(x - \frac{2}{3}\right)^2 + \frac{2}{3}$$

minimum value $= \frac{2}{3}$ when $x = \frac{2}{3}$

minimum turning point $= \left(\frac{2}{3}, \frac{2}{3}\right)$

cuts y-axis at $(0, 2)$

(c) $\quad$ Sketch of $f(x)$

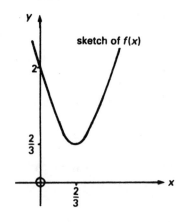

38

25. (a) $U_{r+1} = mU_r + c$ $U_0 = 1, U_1 = -3, u_2 = 21$

$u_1 = mU_0 + c \Rightarrow -3 = m(1) + c$ ①

$U_2 = mU_1 + c \Rightarrow 21 = m(-3) + c$ ②

$\qquad$ ① − ② − 24 = 4m, $\Rightarrow$ $m = -6$

Substitute $m = -6$ in ①

$\qquad\qquad -3 = -6(1) + c$

$\qquad\qquad\qquad 3 = c \qquad\qquad u_2 = -6u_1 + 3$

$\quad m = -6 \quad c = 3 \qquad\qquad$ check $-6(-3) + 3$

$\quad U_{r+1} = -6U_r + 3 \qquad\qquad\qquad = 18 + 3 = 21$

(b) $U_3 = -6u_2 + c$

$\qquad = -6(21) + 3$

$\qquad = -126 + 3 = -123$

$\qquad\underline{U_3 = -123}$

To find U_{-1} use $U_0 = -6U_{-1} + 3$

$\qquad\qquad U_0 = 1 \Rightarrow 1 = -6U_{-1} + 3$

$\qquad\qquad\qquad \Rightarrow -2 = -6U_{-1}$

$\qquad\qquad\qquad \Rightarrow \frac{1}{3} = U^{-1}$

(c) To find $U_r = U_{r+1}$

$\qquad\qquad\qquad U_{r+1} = -6U_r + 3$

$\qquad \Rightarrow \qquad U_r = -6U_r + 3$

$\qquad \Rightarrow \qquad 7U_r = 3$

$\qquad \Rightarrow \qquad U_r = \frac{3}{7}$

check $U_r = \frac{3}{7}$, $U_{r+1} = -6\left(\frac{3}{7}\right) + 3$

$\qquad\qquad\qquad\qquad = \frac{-18}{7} + 3$

$\qquad\qquad\qquad\qquad = \frac{-18}{7} + \frac{21}{7} = \frac{3}{7}$

Gives (U_r, U_{r+1}), $\left(\frac{3}{7}, \frac{3}{7}\right)$, $U_r = U_{r+1}$

WORKED EXAMPLE — TEST PAPER B

1. *(a)* $A = PR^n$
$\qquad\qquad\qquad\qquad$ P = Principal, P = 300
$\qquad\qquad\qquad\qquad$ R = 1 + rate%, R = 1 + 4·5%
$\qquad\qquad\qquad\qquad$ n = number of years, $n = 5$

$\qquad$ $A = 300\,(1·045)^5$
$\qquad\quad\; \fallingdotseq £373·85$

(b) Principal investment = 300
$\qquad$ to double the investment A = 600
$\qquad$ $A = PR^n$
$\qquad$ $600 = 300\,(1·045)^n$
$\qquad\quad 2 = (1·045)^n$
$\qquad$ [This can be found by trial and error method.]

$\qquad$ By log method $\qquad\quad 2 = (1·045)^n$
$\qquad\qquad\qquad\qquad\quad \log 2 = \log 1·045^n$
$\qquad\qquad\qquad\qquad\quad \log 2 = n \log 1·045$
$$\frac{\log 2}{\log 1·045} = n$$
$\qquad\qquad\qquad\qquad\qquad n \fallingdotseq 15·74$

Hence it would take 16 years to double the investment.

2. $f(x) = 2x^3 + mx$
$\quad f'(x) = 6x^2 + m$ (the gradient of the tangent)
$\quad f'(2) = 6(2)^2 + m \qquad$ (Note: $f'(-2) = f'(2)$)
$\quad$ The gradient of the tangent = 0 for stationary values.
$\quad$ Hence $f'(2) = 0$
$\qquad \Rightarrow 24 + m = 0$
$\qquad\qquad\quad m = -24$
$\quad$ So $f(x) = 2x^3 - 24x$
$\quad$ and $f(-1) = 2(-1)^3 - 24(-1)$
$\qquad\qquad\quad = -2 + 24$
$\qquad\qquad\quad = 22$

3. *(a)* Circle $x^2 + y^2 - 2x + 6y + 1 = 0$

General equation of a circle is $x^2 + y^2 + 2gx + 2fy + c = 0$
where centre $= (-g, -f)$ and radius $= \sqrt{(g^2 + f^2 - c)}$
hence centre $= (1, -3)$, radius $= \sqrt{1^2 + 3^2 - 1} = 3$

(b) Centre $(1, -3)$ reflected in y-axis $= (-1, -3)$
$-g = -1$, hence $2g = 2$, f unchanged, r unchanged
equation of circle after reflection in the y-axis is
$x^2 + y^2 + 2x + 6y + 1 = 0$

4. A$(-1, 3)$, B$(2, -1)$, C$(5, 4)$

$m_{AC}\ \dfrac{4-3}{5-(-1)} = \dfrac{1}{6}$ (the gradient of line AC)

If AC is perpendicular to BQ
then $m_{AC} \times m_{BQ} = -1$

$m_{AC} = \dfrac{1}{6}$ hence $m_{BQ} = -6$

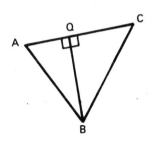

$m_{BQ} = -6$ through B$(2, -1)$
$$y - b = m(x - a)$$
$$y + 1 = -6(x - 2)$$
$$y + 1 = -6x + 12$$
$$y = -6x + 11$$
Equation of BQ is $y + 6x = 11$

5. Graph of $f(x)$

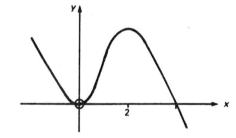

a cubic function $ax^3 + bx + c$

$f'(x)$ will be of form $3ax^2 + b$
which is a quadratic

x	$x < 0$	$x = 0$	$0 < x < 2$	$x = 2$	$x > 2$
$f'(x)$	$-$ve	0	$+$ve	0	$-$ve
	below	on	above	on	below

Plot $f'(x)$ in
relation to
the x-axis

41

x	$x<0$	$x=0$	$0<x<2$	$x=2$	$x>2$
Plotting points in $\Rightarrow$ relation to x-axis	below	on	above	on	below
	Plot	$(0,0)$		$(2,0)$	

graph of $f'(x)$

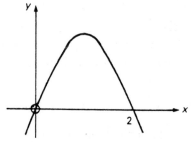

6. $f(x) = x(x^2-4)(x^2-2)(x^2+3) \quad x \in R$

Roots are found when $f(x) = 0$.

Where graph cuts the x-axis i.e. $(x, 0)$

$f(x) = 0 \Rightarrow x = 0, (x^2-4) = 0, (x^2-2) = 0,$ or $(x^2+3) = 0$

$\qquad\qquad x = 0, x = \pm 2 \qquad x = \pm\sqrt{2}, x = \pm\sqrt{-3}, x \in R$

the root of a negative is not real, $x \in R$ [solutions in set of real numbers]

Hence solution set is $\{-2, -\sqrt{2}, 0, \sqrt{2}, 2\}$

7. *(a)*   A(2, −1, 3), B(1, 6, −4)

P divides AB in ratio 2:−3

$\underset{\sim}{a} = \begin{pmatrix} 2 \\ -1 \\ 3 \end{pmatrix}, \underset{\sim}{b} = \begin{pmatrix} 1 \\ 6 \\ -4 \end{pmatrix}$

$\dfrac{\overrightarrow{AP}}{\overrightarrow{PB}} = \dfrac{2}{-3}$

$-3\overrightarrow{AP} = 2\overrightarrow{PB}$

$\Rightarrow -3(\underset{\sim}{p} - \underset{\sim}{a}) = 2(\underset{\sim}{b} - \underset{\sim}{p})$

$\Rightarrow -3\underset{\sim}{p} + 3\underset{\sim}{a} = 2\underset{\sim}{b} - 2\underset{\sim}{p}$

$\Rightarrow \quad 3\underset{\sim}{a} - 2\underset{\sim}{b} = \underset{\sim}{p}$

$\Rightarrow 3\begin{pmatrix} 2 \\ -1 \\ 3 \end{pmatrix} - 2\begin{pmatrix} 1 \\ 6 \\ -4 \end{pmatrix} = \underset{\sim}{p}$

$\begin{pmatrix} 6 \\ -3 \\ 9 \end{pmatrix} - \begin{pmatrix} 2 \\ 12 \\ -8 \end{pmatrix} = \underset{\sim}{p}$

$\begin{pmatrix} 4 \\ -15 \\ 17 \end{pmatrix} = \underset{\sim}{p}$

$\underset{\sim}{p} = \overrightarrow{OP}$ the position vector

Hence P = (4, −15, 17)

42

Alternative Method:

This can also be done by section formula.

$$A(2, -1, 3) \qquad B(1, 6, -4)$$

$$2 \qquad -3$$

$$P = \frac{1}{2 + (-3)} \left[-3 \begin{pmatrix} 2 \\ -1 \\ 3 \end{pmatrix} + 2 \begin{pmatrix} 1 \\ 6 \\ -4 \end{pmatrix} \right]$$

$$= -1 \left[\begin{pmatrix} -6 \\ 3 \\ -9 \end{pmatrix} + \begin{pmatrix} 2 \\ 12 \\ -8 \end{pmatrix} \right]$$

$$= -1 \begin{pmatrix} -4 \\ 15 \\ -17 \end{pmatrix} = (4, -15, 17), \text{ Point P}$$

(b) $\overrightarrow{AB} : \overrightarrow{PB}$

$$\overrightarrow{AB} = \underset{\sim}{b} - \underset{\sim}{a} = \begin{pmatrix} 1 \\ 6 \\ -4 \end{pmatrix} - \begin{pmatrix} 2 \\ -1 \\ 3 \end{pmatrix} = \begin{pmatrix} -1 \\ 7 \\ -7 \end{pmatrix}$$

$$\overrightarrow{PB} = \underset{\sim}{b} - \underset{\sim}{p} = \begin{pmatrix} 1 \\ 6 \\ -4 \end{pmatrix} - \begin{pmatrix} 4 \\ -15 \\ 17 \end{pmatrix} = \begin{pmatrix} -3 \\ 21 \\ -21 \end{pmatrix} = \overrightarrow{PB}, \text{ hence } \overrightarrow{BP} = \begin{pmatrix} 3 \\ -21 \\ 21 \end{pmatrix}$$

$\overrightarrow{AB} : \overrightarrow{PB} = 1 : 3$ and $\overrightarrow{AB} : \overrightarrow{BP} = -1 : 3$

8. (a) $\cos x + \sin x$ in form $k \cos(x - \alpha)$

$k^2 = 1^2 + 1^2 = 2$

$k = \sqrt{2}$

$\tan \alpha = \frac{1}{1} = 1$

$\alpha = 45°$ in the 1st quadrant.

$k = \sqrt{2}, \alpha = 45°$

$k \cos(x - \alpha) = \sqrt{2} \cos(x - 45°)$

(b) $\cos x + \sin x = 1 \quad 0 \leqslant x \leqslant 2\pi$

$45°$ in radians $= \dfrac{45\pi}{180} = \dfrac{\pi}{4}$ radians.

$\cos x + \sin x = 1$

$\Rightarrow \sqrt{2} \cos \left(x - \dfrac{\pi}{4}\right) = 1$

$\Rightarrow \quad \cos \left(x - \dfrac{\pi}{4}\right) = \dfrac{1}{\sqrt{2}}$

cos +ve in quadrant ① and ④

$\cos^{-1}\left(\dfrac{1}{\sqrt{2}}\right) = 45°$ or $315°$

$\Rightarrow \left(x - \dfrac{\pi}{4}\right) = \dfrac{\pi}{4}$ or $\dfrac{7\pi}{4}$

$\Rightarrow x = \dfrac{\pi}{4} + \dfrac{\pi}{4}$ or $x = \dfrac{7\pi}{4} + \dfrac{\pi}{4}$

$x = \dfrac{\pi}{2}$ or 2π

S.S. $\left\{\dfrac{\pi}{2}, 2\pi\right\}$

9. After cleaning $U_{r+1} = 0.25\, U_r + 1 \qquad U_0 = 2$ (Monday before cleaning)

Monday $\quad U_1 = 0.25\,(U_0) + 1 = \dfrac{1}{4}(2) + 1 = 1\dfrac{1}{2}$ or $\dfrac{3}{2}$

Tuesday $\quad U_2 = \dfrac{1}{4}\left(\dfrac{3}{2}\right) + 1 = \dfrac{3}{8} + \dfrac{8}{8} = \dfrac{11}{8} \ \Big|\ 1\dfrac{3}{8}$

Wednesday $\quad U_3 = \dfrac{1}{4}\left(\dfrac{11}{8}\right) + 1 = \dfrac{11}{32} + \dfrac{32}{32} = \dfrac{43}{32} \ \Big|\ 1\dfrac{11}{32}$

Thursday $\quad U_4 = \dfrac{1}{4}\left(\dfrac{43}{32}\right) + 1 = \dfrac{43}{128} + \dfrac{128}{128} = \dfrac{171}{128} \ \Big|\ 1\dfrac{43}{128}$

Friday $\quad U_5 = \dfrac{1}{4}\left(\dfrac{171}{128}\right) + 1 = \dfrac{171}{512} + \dfrac{512}{512} = \dfrac{683}{512} \ \Big|\ 1\dfrac{171}{512}$

Friday after cleaning $\fallingdotseq 1\dfrac{1}{3}$ kg

If this trend continues the litter appears to get closer to $1\dfrac{1}{3}$ kg or $\dfrac{4}{3}$ kg.

If $U_0 = \dfrac{4}{3}$, $U_{r+1} = \dfrac{1}{4}\left(\dfrac{4}{3}\right) + 1$

$= \dfrac{1}{3} + 1$

$= 1\dfrac{1}{3}$ kg

Hence if $\dfrac{4}{3}$ is reached before cleaning there will be $\dfrac{4}{3}$ kg after cleaning.

10. $(1, 2), (a, 4), (b, 1)$

Let points = A, B, C respectively.

If points are collinear then they have equal gradients
i.e. $m_{AB} = m_{BC}$, B lies on line AC

$$m_{AB} = \frac{4-2}{a-1}, m_{BC} = \frac{1-4}{b-a}$$

$$\Rightarrow m_{AB} = \frac{2}{a-1}, m_{BC} = \frac{-3}{b-a}$$

$$\Rightarrow \frac{2}{a-1} = \frac{-3}{b-a}$$

$$\Rightarrow 2(b-a) = -3(a-1)$$

$$\Rightarrow 2b - 2a = -3a + 3$$

$$\Rightarrow a + 2b = 3$$

11. *(a)* $2\cos^2 x + \cos x - 1$

Let $\cos x = x$

$$\Rightarrow 2\cos^2 x + \cos x - 1$$

$$= 2x^2 + x - 1$$

$$\Rightarrow (2x - 1)(x + 1)$$

$$\Rightarrow 2\cos^2 x + \cos x - 1 = (2\cos x - 1)(\cos x + 1)$$

(b) $2\cos^2 x + \cos x - 1 = 0, 0 < x < 360$

$$\Rightarrow (2\cos x - 1)(\cos x + 1) = 0$$

$$\Rightarrow \cos x = \frac{1}{2} \text{ or } \cos x = -1$$

$$\Rightarrow \cos x \text{ lies in quadrants } ① \text{ and } ④ \text{ for } x = \frac{1}{2}$$

$$x = 60, 300°$$

$$\cos x = -1 \Rightarrow x = 180$$

S.S. $\{60°, 180°, 300°\}$

12. $\int_{b}^{2}(3x^2 - 2)dx = 3 \Rightarrow [x^3 - 2x]_{b}^{2} = 3$

$$(2^3 - 2(2)) - (b^3 - 2(b)) = 3$$
$$(8 - 4) - b^3 + 2b = 3$$
$$4 \quad -b^3 + 2b = 3$$
$$1 = b^3 - 2b$$
$$1 = b(b^2 - 2)$$
$$b = -1$$

* Since $b < 2$ Trial and error

Try, $b = 1, 0$ or -1

$-1((-1)^2 - 2)$

$= -1(1 - 2) = -1(-1) = 1$

13.

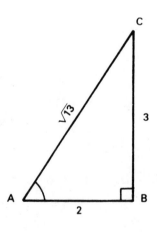

By Pythagoras' theorem

$AC^2 = 3^2 + 2^2 = 13$

$\cos A = \dfrac{2}{\sqrt{13}}$

rationalise denominator $\dfrac{2}{\sqrt{13}} \times \dfrac{\sqrt{13}}{\sqrt{13}} = \dfrac{2\sqrt{13}}{13}$

$\cos A = \dfrac{2\sqrt{13}}{13} = \dfrac{2}{13}\sqrt{13}$

In form $p\sqrt{13} \Rightarrow p = \dfrac{2}{13}$

14. $f(x) = \sin^3 x + \cos^2 x$

$\sin^3 x = (\sin x)^3$

By chain rule method if $f(x) = (\sin x)^3$

$$\text{then } f'(x) = 3(\sin x)^2 \cos x$$
$$= 3 \cos x \sin^2 x$$

and $\cos^2 x = (\cos x)^2$

By chain rule method if $f(x) = (\cos x)^2$

$$\text{then } f'(x) = 2(\cos x)^1(-\sin x)$$
$$= -2 \cos x \sin x$$

Hence $f(x) \qquad = \sin^3 x + \cos^2 x$

$\Rightarrow f'(x) \qquad = 3 \cos x \sin^2 x - 2 \cos x \sin x$

Which factorised $= \sin x \cos x\,(3 \sin x - 2)$

$$f'\left(\frac{\pi}{4}\right) = \sin\frac{\pi}{4} \cos\frac{\pi}{4}\left(3\sin\frac{\pi}{4} - 2\right) \qquad \left(\text{Note: } \sin\frac{\pi}{4} = \cos\frac{\pi}{4}\right.$$

$$= \frac{1}{\sqrt{2}} \cdot \frac{1}{\sqrt{2}}\left(\frac{3}{\sqrt{2}} - 2\right) \qquad\qquad \left. = \frac{1}{\sqrt{2}}\right)$$

$$= \frac{1}{2}\left(\frac{3}{\sqrt{2}} - 2\right)$$

$$= \frac{3}{2\sqrt{2}} - 1$$

$$\text{or } \frac{3 - 2\sqrt{2}}{2\sqrt{2}}$$

15.

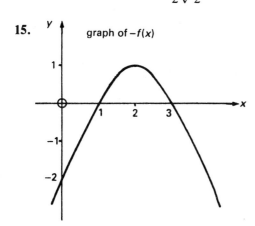

graph of $-f(x)$

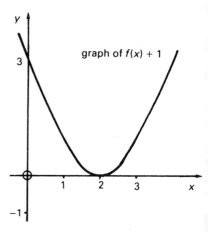

graph of $f(x) + 1$

16. *(a)* $0 \leqslant x \leqslant 360$

$f : x \rightarrow \sin 4x$ cuts x-axis when $\sin 4x = 0$

period $= \dfrac{360}{4} = 90$ 　　　　　$4x = 0, 180$

repeats 4 times 　　　　　$x = 0, 45°$
in 360°

　　　$0, 90 + 0, 180 + 0, 270 + 0, 45, 90 + 45, 180 + 45, 270 + 45$
$x = \{0, 45, 90, 135, 180, 225, 270, 315, 360\}$ 9 times

(b)

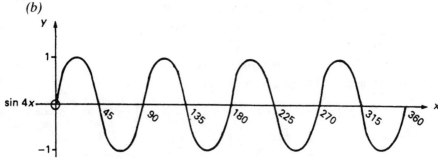

17. *(a)* Line which makes an angle of 45° with the positive direction of the
x-axis has a tangent of 1

　　　　　Hence $m = 1$

Line with gradient $= 1$

Passes through point $(0, -1)$

$y - b = m(x - a)$

$y - (-1) = 1(x - 0)$

$y + 1 = x$

or $y = x - 1$ is required equation.

(b) If a line is perpendicular to another line then the product of their
gradients $= -1$

Hence perpendicular line has gradient $= -1$

Since $1 \times (-1) = -1$

Line with gradient -1 passes through pont $(1, 3)$

$y - b = m(x - a)$

$y - 3 = -1(x - 1)$

$y - 3 = -x + 1$

$y + x = 4$

(c) Perpendicular line has y intercept $(0, 4)$ and since $m = -1$ angle made
is 135°.

18. *(a)* $A = 3 \cos\left(x - \frac{\pi}{6}\right)$

Maximum value = 3

when $\left(x - \frac{\pi}{6}\right) = 0$

since $\cos 0 = 1$

$\Rightarrow x - \frac{\pi}{6} = 0$

$\Rightarrow x = \frac{\pi}{6}$

minimum value = -3

when $x - \frac{\pi}{6} = \pi$

since $\cos \pi = -1$

$\Rightarrow x - \frac{\pi}{6} = \pi$

$x = \pi + \frac{\pi}{6}$

$= \frac{7\pi}{6}$

Maximum value = 3, minimum value = -3

(b) Maximum turning point $\left(\frac{\pi}{6}, 3\right)$

Minimum turning point $\left(\frac{7\pi}{6}, -3\right)$

19. A(1, 2), B(−3, 4), C(5, 6)

$\underset{\sim}{a} = \begin{pmatrix} 1 \\ 2 \end{pmatrix} \quad \underset{\sim}{b} = \begin{pmatrix} -3 \\ 4 \end{pmatrix} \quad \underset{\sim}{c} = \begin{pmatrix} 5 \\ 6 \end{pmatrix}$

We find the position vector of the centroid by:

$\text{centroid} = \frac{1}{3}(\underset{\sim}{a} + \underset{\sim}{b} + \underset{\sim}{c})$

$= \frac{1}{3}\left(\begin{pmatrix} 1 \\ 2 \end{pmatrix} + \begin{pmatrix} -3 \\ 4 \end{pmatrix} + \begin{pmatrix} 5 \\ 6 \end{pmatrix}\right)$

$= \frac{1}{3}\begin{pmatrix} 3 \\ 12 \end{pmatrix} = \begin{pmatrix} 1 \\ 4 \end{pmatrix} = \text{position vector.}$

Hence coordinates of centroid = (1, 4)

20. P(2, −1, 3)

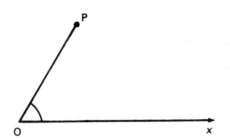

The unit vector parallel to x-axis $= \begin{pmatrix} 1 \\ 0 \\ 0 \end{pmatrix}$

$\underline{p} = \begin{pmatrix} 2 \\ -1 \\ 3 \end{pmatrix}$ $\underline{x} = \begin{pmatrix} 1 \\ 0 \\ 0 \end{pmatrix}$, $|\underline{p}| = 2^2 + 1^2 + 3^2 = \sqrt{14}$, $|\underline{x}| = 1$

$\cos P\hat{O}X = \dfrac{\underline{p} \cdot \underline{x}}{|p||x|} = \dfrac{\begin{pmatrix} 2 \\ -1 \\ 3 \end{pmatrix}\begin{pmatrix} 1 \\ 0 \\ 0 \end{pmatrix}}{(\sqrt{14})(1)} = \dfrac{2 + 0 + 0}{\sqrt{14}} = \dfrac{2}{\sqrt{14}}$

$\cos P\hat{O}X = \dfrac{2}{\sqrt{14}}$

$P\hat{O}X = 57{\cdot}7°$

21.

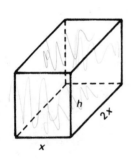

$L \times B \times H = \text{Volume}, \quad \text{Volume} = 576 \text{ cm}^3$

$2x \times x \times h = 576$

$2x^2h = 576$

$h = \dfrac{576}{2x^2}$

Surface Area $= 2 \times x(2x) + 2 \times xh + 2 \times 2xh$

$\qquad = 4x^2 + 2xh + 4xh \Rightarrow \text{S.A.} = 4x^2 + 6xh$

$h = \dfrac{576}{2x^2} \Rightarrow \text{S.A.} = 4x^2 + 6x\left(\dfrac{576}{2x^2}\right)$

$\qquad\qquad \text{S.A.} = 4x^2 + \dfrac{1728}{x}$

Let function for surface area = $S(x)$

$$S(x) = 4x^2 + \frac{1728}{x} = 4x^2 + 1728x^{-1}$$

Use calculus for minimum T.P., $\qquad S'(x) = 8x - 1728x^{-2}$

$$= m \tan = 0 \text{ (S.V.)}$$

find stationary value (S.V.)

$\qquad m \tan = 0$

$$\Rightarrow 8x = \frac{1728}{x^2}$$

at $x = 6$ $S(x)$ has T.P.

check if minimum $\quad x < 6, x > 6$

$$\Rightarrow 8x^3 = 1728$$

$$x^3 = 216$$

$$S'(5) = 8(5) - \frac{1728}{5^2} -ve \searrow$$

$$x = 6$$

$$S'(7) = 8(7) - \frac{1728}{7^2} +ve \nearrow$$

$\qquad\qquad\qquad S'(5) \qquad S'(6) \qquad S'(7)$

$\qquad\qquad\qquad \searrow \qquad\quad \rightarrow \qquad\quad \nearrow$

Hence $x = 6$ gives a minimum value for the function $S(x)$
Length = 12 cm. Breadth = 6 cm. Height = 8 cm.

22.
$\qquad\qquad\qquad 1 \cdot 2^x = 12$ $\qquad$ (This can be solved by trial and

Log method: $\quad \log 1 \cdot 2^x = \log 12$ $\qquad$ error on the calculator.)

$\qquad\qquad x \log 1 \cdot 2 = \log 12$

$$x = \frac{\log 12}{\log 1 \cdot 2} \approx 13 \cdot 63, \text{ Test } 1 \cdot 2^{13 \cdot 63} \approx 12$$

23. $U_{r+1} = mU_r + c \quad U_0 = -1, U_1 = 7, U_2 = -9$

(a) $U_1 = m(U_0) + c \Rightarrow 7 = m(-1) + c \;①$

$\qquad U_2 = m(U_1) + c \Rightarrow -9 = m(7) + c \;②$

$\qquad\qquad ② - ① - 16 = 8m \Rightarrow m = -2$

$\qquad$ Substitute $m = -2$ in $①$ $7 = -2(-1) + c$

$$7 = 2 + c \Rightarrow c = 5$$

Recurrence relation is $U_{r+1} = -2U_r + 5$

(b) $U_3 = -2U_2 + 5$

$\quad\quad = -2(-9) + 5$

$\quad U_3 = 23$

U_{-1} is found by using U_0

$U_0 = -2U_{-1} + 5$

$-1 = -2U_{-1} + 5$

$-6 = -2U_{-1}$

$\dfrac{-6}{-2} = U_{-1}$

$\quad 3 = U_{-1}$

$U_{-1} = 3$

(c) $\quad U_r = U_{r+1}$

$\Rightarrow \; U_r = -2U_r + 5$

$\Rightarrow 3U_r = 5$

$\quad\quad U_r = \dfrac{5}{3}$

Test $-2\left(\dfrac{5}{3}\right) + 5 = \dfrac{5}{3}$

24. $\quad g(x) = 3 - x^2$

$\quad\quad f(x) = 1 - 2x$

$\quad g(f(x)) = g(1 - 2x) = 3 - (1 - 2x)^2$

$\quad\quad\quad\quad\quad\quad\quad = 3 - (1 + 4x^2 - 4x)$

$\quad\quad\quad\quad\quad\quad\quad = 3 - 1 - 4x^2 + 4x$

$\quad\quad\quad\quad\quad\quad\quad = 2 + 4x - 4x^2$

$\quad\quad g(f(x)) = 2(1 + 2x - 2x^2)$

25. *(a)* $f(x) = 3x^2 - 2x + 5$

$\quad\quad a = 3, b = -2, c = 5$

$\quad\quad\quad b^2 - 4ac$

$\quad\quad (-2)^2 - 4(3)(5)$

$\quad\quad\quad 4 \; -60 = -56$

$\quad\quad\quad \sqrt{-56} \in R$

using the discriminant $b^2 - 4ac$

If $b^2 - 4ac < 0$

$f(x)$ has no real roots.

$\quad f(x)$ has no real roots.

52

(b) <u>Completing the square</u>

$3x^2 - 2x + 5$

$$= 3\left(x^2 - \frac{2}{3}x + \frac{5}{3}\right)$$

$$= 3\left[\left(x - \frac{2}{6}\right)^2 - \left(\frac{2}{6}\right)^2 + \frac{5}{3}\right]$$

$$= 3\left[\left(x - \frac{1}{3}\right)^2 - \frac{1}{9} + \frac{5}{3}\right]$$

$$= 3\left[\left(x - \frac{1}{3}\right)^2 + \frac{14}{9}\right]$$

$$= 3\left(x - \frac{1}{3}\right)^2 + \frac{14}{3}$$

$\Rightarrow$ minimum value $= \frac{14}{3}$ when $x = \frac{1}{3}$

Since $\left(x - \frac{1}{3}\right)^2$ is never negative.

Sketch of $3x^2 - 2x + 5$ cuts y-axis at $(0, 5)$

Minimum T.P. $\left(\frac{1}{3}, \frac{14}{3}\right)$.

(c) sketch of $f(x)$

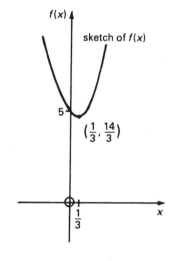

WORKED EXAMPLE — TEST PAPER C

1. $\int_a^6 (2x - 3)\,dx = 20 \Rightarrow [x^2 - 3x]_a^6 = 20$

$$(6^2 - 3(6)) - (a^2 - 3a) = 20$$
$$36 - 18 - a^2 + 3a = 20$$
$$18 - a^2 + 3a = 20$$
$$3a - a^2 = 2$$
$$a(3 - a) = 2$$

By trial and error since $a < 6$
try 3, 2, 1, etc.
try $a = 3, 3(3 - 3) \neq 2$
$a = 1, 1(3 - 1) = 2$
$1(2) = 2$
hence $a = 1$

2. $\int_a^0 (f_1(x) - f_2(x))\,dx + \int_0^b (f_2(x) - f_1(x))\,dx = \text{(upper} - \text{lower)}$
$x < 0$ line is upper,　$x > 0$ curve is upper
　　line $= f_1(x)$　　　　curve $= f_2(x)$

3. $f : x \rightarrow \sin 3x$　　$0 \leqslant x < 360$

period $= \dfrac{360}{3} = 120$

curve repeats 3 times
in interval 0 to 360

cuts x-axis when $\sin 3x = 0$
$\Rightarrow 3x = 0, 180, 360, 540,$ etc.
$x = \{0, 60, 120, 180, 240, 300\}$
$\sin 3x$ meets the x-axis 6 times.

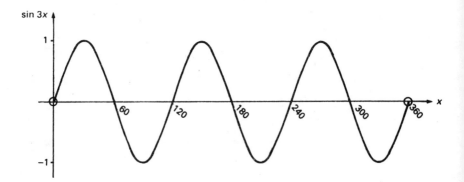

4. $x^2 + y^2 - 8x + 6y + 21 = 0$　　　$2g = -8, -g = 4$
　　(radius remains the same)　　　$2f = 6, -f = -3$　centre $(-g, -f)$
　　　　centre $(4, -3)$ reflected in y-axis centre $= (-4, -3)$
　　　　equation is $x^2 + y^2 + 8x + 6y + 21 = 0$

54

5. $0.3^n < 0.02$, find $0.3^n = 0.02$ * This can be found by trial and error.

Log Method.

$$0.3^n = 0.02$$
$$\log 0.3^n = \log 0.02$$
$$n \log 0.3 = \log 0.02$$
$$n = \frac{\log 0.02}{\log 0.3}$$

$$n \geqslant 3.25$$

$n \approx 3.25$, $0.3^{3.25} \approx 0.0198$ etc.
$0.3^n < 0.02 \Rightarrow n > 3.25$
Test $0.3^{3.3} = 0.0189$

6. $f(x) = 2 \sin 3x$
$f'(x) = 6 \cos 3x$,

$$f'\left(\frac{\pi}{4}\right) = 6 \cos 3\left(\frac{\pi}{4}\right)$$

$$= 6 \times \frac{-1}{\sqrt{2}} = \frac{-6}{\sqrt{2}} = -3\sqrt{2}$$

7. (a) P(1, 2), Q(6, 3), R(5, −2)

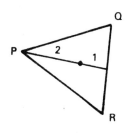

Let mid point of QR = L

$$L = \left(\frac{6+5}{2}, \frac{3-2}{2}\right)$$

$$L = \left(\frac{11}{2}, \frac{1}{2}\right)$$

Let C = centroid
PC : CL = 2:1

$P \xrightarrow{\quad 2 \quad} \underset{C}{\bullet} \xrightarrow{\quad 1 \quad} L$

$$C = \frac{1}{3}\left[\binom{1}{2} + \binom{11}{1}\right], = \frac{1}{3}\binom{12}{3} = (4, 1)$$

<u>Alternative Method:</u>

$$\overrightarrow{OC} = \tfrac{1}{3}(\underset{\sim}{p} + \underset{\sim}{q} + \underset{\sim}{r}) = \left(\frac{1+6+5}{3}, \frac{2+3+(-2)}{3}\right) = (4, 1) = C$$

(b) P(1, 2), Q(6, 3), R(5, –2)

$$M = \tfrac{1}{2}(\underset{\sim}{p} + \underset{\sim}{r}) = \tfrac{1}{2}\binom{6}{0} = (3, 0),$$

$$N = \tfrac{1}{2}(\underset{\sim}{q} + \underset{\sim}{p}) = \tfrac{1}{2}\binom{7}{5} = \left(\tfrac{7}{2}, \tfrac{5}{2}\right)$$

$$\overrightarrow{QC} = \underset{\sim}{c} - \underset{\sim}{q} \qquad \underset{\sim}{c}\binom{4}{1}, \underset{\sim}{q}\binom{6}{3}$$

$$= \binom{-2}{-2}$$

$$\overrightarrow{QM} = \underset{\sim}{m} - \underset{\sim}{q}, \underset{\sim}{m}\binom{3}{0}, \underset{\sim}{q}\binom{6}{3}$$

$$= \binom{-3}{-3}$$

$$\overrightarrow{QC}:\overrightarrow{QM}$$
$$= 2:3$$

$$\overrightarrow{RC} = \underset{\sim}{c} - \underset{\sim}{r}, \underset{\sim}{c}\binom{4}{1}, \underset{\sim}{r}\binom{5}{-2} \qquad \overrightarrow{CN} = \underset{\sim}{n} - \underset{\sim}{c}, \underset{\sim}{n} = \binom{\tfrac{7}{2}}{\tfrac{5}{2}}, \underset{\sim}{c} = \binom{4}{1}$$

$$\underset{\sim}{c} - \underset{\sim}{r} = \binom{4-5}{1-(-2)} \qquad\qquad \underset{\sim}{n} - \underset{\sim}{c} = \binom{\tfrac{7}{2}-\tfrac{8}{2}}{\tfrac{5}{2}-\tfrac{2}{2}}$$

$$= \binom{-1}{3} \qquad\qquad = \binom{-\tfrac{1}{2}}{\tfrac{3}{2}}$$

Ratio $\overrightarrow{RC}:\overrightarrow{CN}$ = 2:1

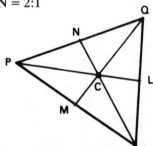

8. $f(x) = x^3 - 3x^2 - 4x + 12$

$$
\begin{array}{c|rrrr}
 & 1 & -3 & -4 & +12 \\
-2 & & -2 & 10 & -12 \\
\hline
 & 1 & -5 & +6 & 0
\end{array}
$$

Try factors of -12
$\{\pm 1, \pm 2, \pm 3, \pm 4, \pm 6\}$
If $f(x) \div (x + 2)$ has remainder zero
then $(x + 2)$ is a factor

Hence $f(x) = (x + 2)(x^2 - 5x + 6)$
$\qquad\quad = (x + 2)(x - 3)(x - 2)$

9. By Pythagoras' theorem the 3rd side of the triangle

$$= \sqrt{(3^2 + 7^2)} = \sqrt{58}$$

$$\cos A = \frac{7}{\sqrt{58}}$$

$\cos 2A = 2 \cos^2 A - 1$

$$= 2\left(\frac{7}{\sqrt{58}}\right)^2 - 1$$

$$= 2\left(\frac{49}{58}\right) - 1$$

$$= \frac{98}{58} - \frac{58}{58} = \frac{40}{58} = \frac{20}{29}$$

$$\cos 2A = \frac{20}{29}$$

10. $f(x) = (2x^2 - x)^5$

By chain rule method
$f'(x) = 5(2x^2 - x)^4(4x - 1)$
$f'(x) = 5(4x - 1)(2x^2 - x)^4$

11

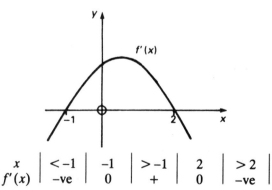

x	< -1	-1	> -1	2	> 2
$f'(x)$	$-$ve	0	$+$	0	$-$ve

$f(x)$ has general shape of ax^3, a negative
$f'(x)$ has general shape of $-3x^2$ ⌢

Plot $(-1, 0)(2, 0)$

12. $C = 3 \cos\left(x + \frac{\pi}{3}\right)$ maximum value = 3 when $x + \frac{\pi}{3} = 0$ or 2π

minimum value = -3 when $x + \frac{\pi}{3} = \pi$

min, $x + \frac{\pi}{3} = \pi, x = \pi - \frac{\pi}{3}, x = \frac{2\pi}{3}$ max, $x = -\frac{\pi}{3}$ or $2\pi - \frac{\pi}{3} = \frac{5\pi}{3}$

$x = \frac{5\pi}{3}$ $y = 3$, $x = \frac{2\pi}{3}$ $y = -3$

13. $f(x) = x(x + 2)(x^2 - 5)(x^2 + 1)(x^2 - 9) = 0$
$\Rightarrow x = 0, x = -2, x = \pm\sqrt{5}, x = \pm 3$
Note $(x^2 + 1) \neq 0$ since $\sqrt{-ve} \notin R$.
S.S. $\{-3, -\sqrt{5}, -2, 0, \sqrt{5}, 3\}$

14. (a)

$$\underset{\sim}{a}\begin{pmatrix} 1 \\ -2 \\ 4 \end{pmatrix} \overset{}{\underset{2\quad 1}{\times}} \underset{\sim}{b}\begin{pmatrix} -2 \\ 4 \\ 1 \end{pmatrix} \qquad \overrightarrow{AP}:\overrightarrow{PB} = 2:1$$

By section formula method

$$P = \tfrac{1}{3}(\underset{\sim}{a} + 2\underset{\sim}{b}) = \tfrac{1}{3}\left[\begin{pmatrix} 1 \\ -2 \\ 4 \end{pmatrix} + \begin{pmatrix} -4 \\ 8 \\ 2 \end{pmatrix}\right] = \tfrac{1}{3}\begin{pmatrix} -3 \\ 6 \\ 6 \end{pmatrix}$$

$P = (-1, 2, 2)$

(b) $\overrightarrow{BP} = -\overrightarrow{PB}$

hence $\overrightarrow{AP}:\overrightarrow{BP}$

$= \overrightarrow{AP}:-\overrightarrow{PB}$

$= 2:-1$

$\overrightarrow{AP}:\overrightarrow{BP} = 2:-1$

* For Alternative Method see *
Paper A Question 11 and Paper B Question 7

15. L(2, 4), M(−1, −2), N(3, 7)

$$m_{MN} = \frac{7-(-2)}{3-(-1)} = \frac{9}{4}$$

$m_{MN} \cdot m_{LQ} = -1$ for LQ ⊓ to MN

Hence $m_{LQ} = -\frac{4}{9}$ through point L(2, 4)

$$y - 4 = -\frac{4}{9}(x - 2)$$

$$9y - 36 = -4x + 8$$

$$9y + 4x = 44 = \text{equation of LQ.}$$

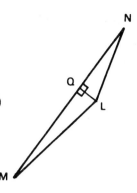

16. *(a)* $x^2 + y^2 - 6x + 8y + 9 = 0$, $C(-g, -f)$, $r = \sqrt{g^2 + f^2 - 9}$

$2g = -6, -g = 3,$

$2f = 8, -f = -4$

Centre $= (3, -4)$, radius $= \sqrt{(3^2 + 4^2 - 9)} = \sqrt{25 - 9} = \sqrt{16}$

Radius $= 4$

(b) After reflection in the x-axis $\quad$ ✓ $\quad (3, -4) \rightarrow (3, 4)$

radius unchanged $\quad$ ✓

equation is $x^2 + y^2 - 6x - 8y + 9 = 0$

* For Alternative Method see *
Paper A Question 15 and Paper B Question 3

17. *(a)* $f(x) = 4x^3 + mx$

$f'(x) = 12x^2 + m = m \tan = 0$ $\qquad$ $m \tan$ is the gradient of the tangent to the curve.

$x = \pm\frac{\sqrt{3}}{2}$ $\quad 12\left(\frac{\sqrt{3}}{2}\right)^2 + m = 0$ $\qquad$ $m \tan = 0$ for stationary values.

$$12\left(\frac{3}{4}\right) + m = 0$$

$$9 + m = 0, \Rightarrow m = -9$$

(b) Hence $f(x) = 4x^3 - 9x$

and $f(-2) = 4(-2)^3 - 9(-2)$

$\qquad \Rightarrow 4(-8) + 18$

$\qquad \Rightarrow -32 + 18$

$\qquad f(-2) = -14$

18. $(2\sqrt{3} - 5\sqrt{2})^2 = (2\sqrt{3} - 5\sqrt{2})(2\sqrt{3} - 5\sqrt{2})$

$\qquad\qquad\qquad\quad = 2\sqrt{3}(2\sqrt{3} - 5\sqrt{2}) - 5\sqrt{2}(2\sqrt{3} - 5\sqrt{2})$

$\qquad\qquad\qquad\quad = 12 - 10\sqrt{6} - 10\sqrt{6} + 50$

$\qquad\qquad\qquad\quad = 62 - 20\sqrt{6}$

19. *(a)* $A = PR^n, R = 1 + \dfrac{6 \cdot 5}{100}, P = 500, n = 10, r = 6 \cdot 5\%$

$\qquad 500(1 \cdot 065)^{10} = £938.57$ amount after 10 years.

(b) If interest is added $\frac{1}{2}$ yearly $n = 20, r = 3 \cdot 25$

$\qquad$ Then $PR^n = 500(1 \cdot 0325)^{20} = £947.92$

$\qquad$ Difference: $947 \cdot 92 - 938 \cdot 57 = £9.35$.

20.

$$\underset{\sim}{p} = \begin{pmatrix} -2 \\ 1 \\ 3 \end{pmatrix}, \underset{\sim}{y} = \begin{pmatrix} 0 \\ 1 \\ 0 \end{pmatrix}$$

$$\cos P\hat{O}Y = \frac{\underset{\sim}{p} \cdot \underset{\sim}{y}}{|\underset{\sim}{p}||\underset{\sim}{y}|} = \frac{\begin{pmatrix} -2 \\ 1 \\ 3 \end{pmatrix} \cdot \begin{pmatrix} 0 \\ 1 \\ 0 \end{pmatrix}}{\left|\begin{pmatrix} -2 \\ 1 \\ 3 \end{pmatrix}\right|\left|\begin{pmatrix} 0 \\ 1 \\ 0 \end{pmatrix}\right|} = \frac{1}{\sqrt{14}\sqrt{1}} = \frac{1}{\sqrt{14}}$$

$$\cos P\hat{O}Y = \frac{1}{\sqrt{14}}, P\hat{O}Y = 74 \cdot 5°$$

21. $5^n > 9^{10}$

This can be solved by trial and error or by log method.

Log method:

$\log 5^n = \log 9^{10}$

$n \log 5 = 10 \log 9$

$n = \dfrac{10 \log 9}{\log 5}$ 　　$n \doteqdot 13\cdot65,$ 　　$5^n > 9^{10}$

$\Rightarrow n > 13\cdot65$

least $+$ve integer $= 14$

$n = 14$

22. A point on y-axis $= (0, y), (7, 2), (-1, 3)$

　　　　　　　　　　　　　A　　B　　C

given 　$|\overrightarrow{AB}|^2 = |\overrightarrow{AC}|^2$

$\Rightarrow |\underset{\sim}{b} - \underset{\sim}{a}|^2 = |\underset{\sim}{c} - \underset{\sim}{a}|^2$

$\underset{\sim}{a} = \begin{pmatrix} 0 \\ y \end{pmatrix}, \underset{\sim}{b} = \begin{pmatrix} 7 \\ 2 \end{pmatrix}, \underset{\sim}{c} = \begin{pmatrix} -1 \\ 3 \end{pmatrix}$

$|\underset{\sim}{b} - \underset{\sim}{a}|^2 = |\underset{\sim}{c} - \underset{\sim}{a}|^2$

$\Rightarrow 7^2 + (2 - y)^2 = (-1)^2 + (3 - y)^2$

$49 + 4 - 4y + y^2 = 1 + 9 - 6y + y^2$

$53 - 4y + y^2 = 10 - 6y + y^2$

$53 - 4y = 10 - 6y$

$2y = -43, y = -21\tfrac{1}{2}$

Point on y-axis $(0, -21\tfrac{1}{2})$

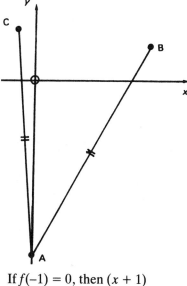

23. 　　$f(x) = x^3 - 5x^2 - x + d$

If $f(-1) = 0$, then $(x + 1)$ is a factor.

```
       1  -5  -1   d
 -1       -1   6  -5
   _____
       1  -6   5  -5 + d,  d - 5 = 0, d = 5
```

Hence $f(x) = (x + 1)(x^2 - 6x + 5)$

$f(x) = (x + 1)(x - 1)(x - 5)$ fully factorised.

61

24. *(a)*

$$0 \leqslant x \leqslant 360$$

$2 \sin x - 5 \cos x$ $k = \sqrt{2^2 + (-5)^2}$

$\tan \theta = \dfrac{2}{-5}$ $\begin{array}{l}\sin + \\ \cos -\text{ve}\end{array}$ $= \sqrt{29}$

in quadrant 2

$\theta = 158 \cdot 2°$

In form $k \cos(x - \theta)$

gives $\sqrt{29} \cos(x - 158 \cdot 2°)$

Maximum value $= -\sqrt{29}$ when $(x - 158 \cdot 2) = 0$ or 360

$\Rightarrow x = 158 \cdot 2$ Point $(158 \cdot 2, \sqrt{29})$

Minimum value $= -\sqrt{29}$ when $(x - 158 \cdot 2) = 180$

$\Rightarrow x = 180 + 158 \cdot 2$

$= 338 \cdot 2°$ Point $(338 \cdot 2, -\sqrt{29})$

Cuts y-axis when $x = 0$, $\cos(0 - 158 \cdot 2) = \cos(-158 \cdot 2)$

$= \cos(158 \cdot 2) = -0 \cdot 928$

$\sqrt{29} \cos(158 \cdot 2) = -5$ Point $(0, -5)$

Cuts x-axis at $\cos 90$ and $\cos 270$

when $(x - 158 \cdot 2) = 90$ when $x - 158 \cdot 2 = 270$

$x = 248 \cdot 2$ $x = 270 + 158 \cdot 2 = 428 \cdot 2$

Point $(248 \cdot 2, 0)$ $428 \cdot 2 - 360 = 68 \cdot 2$

Point $(68 \cdot 2, 0)$

y intercept	zero	maximum	zero	minimum
Points $(0, -5)$,	$(68 \cdot 2, 0)$,	$(158 \cdot 2, \sqrt{29})$,	$(248 \cdot 2, 0)$,	$(338 \cdot 2 - \sqrt{29})$, $(360, -5)$

(b)

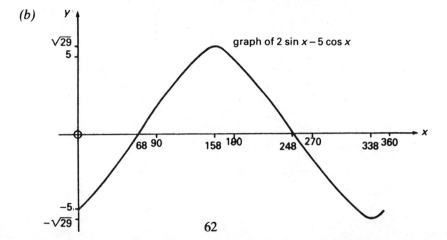

graph of $2 \sin x - 5 \cos x$

62

25. (a) $U_{r+1} = mU_r + c$ $U_0 = 3, U_1 = 2, U_2 = 4$

$U_1 = mU_0 + c \Rightarrow 2 = 3m + c$ ①

$U_2 = mU_1 + c \Rightarrow 4 = 2m + c$ ②

$$② - ① \qquad 2 = -m$$

substitute $m = -2$ in ①, $2 = 3(-2) + c$

$$\Rightarrow 2 = -6 + c$$

$$c = 8$$

$m = -2, c = 8$

$U_{r+1} = -2U_r + 8$

(b) $U_3 = -2U_2 + 8$

$= -2(4) + 8 = 0$

U_{-1} is found by using $U_0 = -2U_{-1} + 8, U_0 = 3$

$$3 = -2U_{-1} + 8$$

$$-5 = -2U_{-1}$$

$$\frac{5}{2} = U_{-1}$$

Hence $U_3 = 0, U_{-1} = \frac{5}{2}$

(c) $U_r = U_{r+1} \Rightarrow U_r = -2U_r + 8$

$$\Rightarrow 3U_r = 8$$

$$U_r = \frac{8}{3} \qquad \text{Check } -2\left(\frac{8}{3}\right) + 8$$

$$= \frac{-16}{3} + \frac{24}{3} = \frac{8}{3}$$

$(U_r, U_{r+1}) = \left(\frac{8}{3}, \frac{8}{3}\right)$

WORKED EXAMPLE — TEST PAPER D

1. $(5 + 2\sqrt{3})^2 = (5 + 2\sqrt{3})(5 + 2\sqrt{3})$

$$= 5(5 + 2\sqrt{3}) + 2\sqrt{3}(5 + 2\sqrt{3})$$
$$= 25 + 10\sqrt{3} + 10\sqrt{3} + 12$$
$$= 37 + 20\sqrt{3}$$

2. $\cos 4x$ $\qquad 0 \leqslant x \leqslant 180$ $\qquad \cos 4x = 0 \Rightarrow 4x = 90, 270$

$\text{period} = \dfrac{360}{4} = 90$ $\qquad\qquad 4x = 90, x = 22\frac{1}{2}, 112\frac{1}{2}$

$\qquad\qquad\qquad\qquad\qquad\qquad\qquad 4x = 270, x = 67\frac{1}{2}, 157\frac{1}{2}$

repeats twice in 180° $\qquad$ cuts at $x = \{22\frac{1}{2}, 67\frac{1}{2}, 112\frac{1}{2}, 157\frac{1}{2}\}$

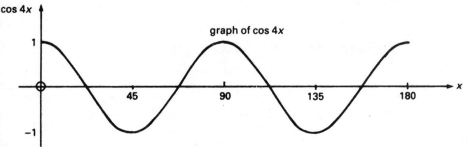

crosses the x-axis in 4 places

3. A(2, 1), B(a, 5), C(b, –7) in component form.

$$\underset{\sim}{a} = \begin{pmatrix} 2 \\ 1 \end{pmatrix}, \underset{\sim}{b} = \begin{pmatrix} a \\ 5 \end{pmatrix}, \underset{\sim}{c} = \begin{pmatrix} b \\ -7 \end{pmatrix}$$

If points are collinear B is a shared point and $\overrightarrow{AB}$ is parallel to $\overrightarrow{BC}$.

$$\Rightarrow \underset{\sim}{b} - \underset{\sim}{a} = k(\underset{\sim}{c} - \underset{\sim}{b})$$

$$\underset{\sim}{b} - \underset{\sim}{a} = \begin{pmatrix} a - 2 \\ 4 \end{pmatrix}, \underset{\sim}{c} - \underset{\sim}{b} = \begin{pmatrix} b - a \\ -12 \end{pmatrix}$$

If collinear ratio is $4{:}{-}12$

$$= 1{:}{-}3$$

$$\Rightarrow -3(a - 2) = b - a$$
$$-3a + 6 = b - a$$
$$\Rightarrow 6 = 2a + b$$

64

Alternative method:

A(2, 1), B(a, 5), C(b, -7)

$m_{AB} = \dfrac{5-1}{a-2}$, $m_{BC} = \dfrac{5-(-7)}{a-b}$

$\dfrac{4}{a-2} = \dfrac{12}{a-b}$ if parallel

$4a - 4b = 12a - 24$

$\Rightarrow 24 = 8a + 4b \Rightarrow 6 = 2a + b$ as given.

4. General equation of circle $= x^2 + y^2 + 2gx + 2fy + c = 0$,

   centre $= (-g, -f)$, $\quad r = \sqrt{(g^2 + f^2 - c)}$

   given equation $= x^2 + y^2 - 8x + 6y + 21 = 0$

   $2g = -8, \Rightarrow -g = 4, 2f = 6 \Rightarrow -f = -3,$

   radius $= \sqrt{(4^2 + 3^2 - 21)} = \sqrt{4} = 2$

   Centre $= (4, -3)$ reflected in $y = -x$ becomes $(3, -4)$, $r = 2$.

   Hence equation of circle after reflection in $y = -x$:

   $\quad x^2 + y^2 - 6x + 8y + 21 = 0$

5. By Pythagoras' theorem AB $= \sqrt{(2\sqrt{2})^2 + 1^2} = 3$

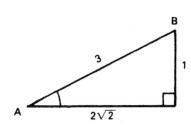

$\cos A = \dfrac{2\sqrt{2}}{3}$

$\sin A = \dfrac{1}{3}$

$\sin 2A = 2 \sin A \cos A$

$\sin 2A = 2\left(\dfrac{1}{3}\right)\left(\dfrac{2\sqrt{2}}{3}\right) = \dfrac{4\sqrt{2}}{9}$

$\cos 2A = 2\cos^2 A - 1 = 2\left(\dfrac{2\sqrt{2}}{3}\right)^2 - 1$

$= 2\left(\dfrac{8}{9}\right) - 1$

$\cos 2A = \dfrac{16}{9} - \dfrac{9}{9} = \dfrac{7}{9}$

$\cos 2A = \dfrac{7}{9}$, $\quad \sin 2A = \dfrac{4\sqrt{2}}{9}$

6. $0 \cdot 5^n < 0 \cdot 1$, find $0 \cdot 5^n = 0 \cdot 1$ (can be found by trial and error)

Log method $\log 0 \cdot 5^n = \log 0 \cdot 1$

$\qquad n \log 0 \cdot 5 = \log 0 \cdot 1$

$\qquad n \qquad = \dfrac{\log 0 \cdot 1}{\log 0 \cdot 5}$

$\qquad\qquad n \fallingdotseq 3 \cdot 32$

$\qquad$ check $n = 3 \cdot 32$ $\quad 0 \cdot 5^{3 \cdot 32}$ is greater than $0 \cdot 1$

$\qquad$ hence $n > 3 \cdot 32$

7. $f(x) = (3x^2 - 2x)^4$

By chain rule method.

$f'(x) = 4(3x^2 - 2x)^3(6x - 2)$

$\qquad = 4(6x - 2)(3x^2 - 2x)^3$

$f'(x) = 4(6x - 2)(3x^2 - 2x)^3$

$f'(-1) = 4(6(-1) - 2)(3(-1)^2 - 2(-1))^3$

$f'(-1) = 4(-6 - 2)(3 + 2)^3$

$\qquad\quad 4(-8)(5)^3$

$f'(-1) = -32 \times 125 = -4000$

8. $D = 2 \cos\left(x - \dfrac{\pi}{2}\right)$ minimum value $= -2$ when $x - \dfrac{\pi}{2} = \pi$

$$\Rightarrow \quad x = \pi + \dfrac{\pi}{2} = \dfrac{3\pi}{2}$$

$\qquad$ maximum value $= 2$

$\qquad$ when $\left(x - \dfrac{\pi}{2}\right) = 0$ or 2π,

$\qquad x = \dfrac{\pi}{2}$ or $\dfrac{5\pi}{2}$ (too large)

$\qquad (x, y) = \left(\dfrac{\pi}{2}, 2\right), \left(\dfrac{3\pi}{2}, -2\right)$

9. $f(x) = ax^3, \quad a$ +ve

$f'(x) = 3ax^2 \Rightarrow$ shape $\smile$

x	<-3	-3	>-3	0	>0
$f'(x)$	$+$	0	$-$	0	$+$
Plot	above	on	below	on	above

Points in relation to x-axis.

Plot $(-3, 0), (0, 0)$

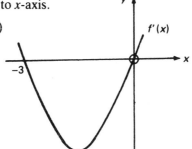

10. *(a)*

$\underset{\sim}{a} \begin{pmatrix} 2 \\ -2 \\ 5 \end{pmatrix} \quad \underset{\sim}{b} \begin{pmatrix} -2 \\ 2 \\ -3 \end{pmatrix}$

$\underset{1}{\times}_{3}$

$P = \frac{1}{4}(3\underset{\sim}{a} + \underset{\sim}{b}) = \frac{1}{4}\left[\begin{pmatrix} 6 \\ -6 \\ 15 \end{pmatrix} + \begin{pmatrix} -2 \\ 2 \\ -3 \end{pmatrix}\right] = \frac{1}{4}\begin{pmatrix} 4 \\ -4 \\ 12 \end{pmatrix}$

$P = (1, -1, 3)$

(b)

$\overrightarrow{AB} = \underset{\sim}{b} - \underset{\sim}{a} = \begin{pmatrix} -4 \\ 4 \\ -8 \end{pmatrix}$

$\overrightarrow{PB} = \underset{\sim}{b} - \underset{\sim}{p} = \begin{pmatrix} -3 \\ 3 \\ -6 \end{pmatrix}$

ratio $\overrightarrow{AB}:\overrightarrow{PB}$

$= 4 : 3$

A —— 1 —— P —— 3 —— B

11. *(a)* $3\cos^2 x - 8\cos x + 4$

$= (3\cos x - 2)(\cos x - 2)$

Let $\cos x = x$,

$3x^2 - 8x + 4$

$= (3x - 2)(x - 2)$

(b) $\cos x = \frac{2}{3}$ or $\cos x = 2$ (not possible)

$x = 48 \cdot 2, 360 - 48 \cdot 2,$

s.s. $\{48 \cdot 2, 311 \cdot 8\}$

12. Let given points be P, Q, R respectively.

(2, 5)   (4, 1)   (8, –3)
  P   Q   R

Mid point QR $= \left(\dfrac{4+8}{2}, \dfrac{1+(-3)}{2} \right)$

$= (6, -1), M = (6, -1)$

Let C = centroid   PC:CM = 2:1   P $\underline{\quad\quad^2\quad\quad \bullet_{C} \quad^1\quad}$ M

$\underset{2}{\underline{p}} \begin{pmatrix} 2 \\ 5 \end{pmatrix} \underset{1}{\overset{\underline{m} \begin{pmatrix} 6 \\ -1 \end{pmatrix}}{\bowtie}}$

$C = \tfrac{1}{3}\left[\begin{pmatrix} 12 \\ -2 \end{pmatrix} + \begin{pmatrix} 2 \\ 5 \end{pmatrix} \right] = \tfrac{1}{3}\begin{pmatrix} 14 \\ 3 \end{pmatrix}$ $C = \left(\tfrac{14}{3}, 1 \right)$

<u>Alternative Method</u>

$\overrightarrow{OC} = \tfrac{1}{3}(\underline{p} + \underline{q} + \underline{r})$

$= \tfrac{1}{3}\left[\begin{pmatrix} 2 \\ 5 \end{pmatrix} + \begin{pmatrix} 4 \\ 1 \end{pmatrix} + \begin{pmatrix} 8 \\ -3 \end{pmatrix} \right] = \tfrac{1}{3}\begin{pmatrix} 14 \\ 3 \end{pmatrix}$

$C = \left(\tfrac{14}{3}, 1 \right)$

13. $\displaystyle\int_a^2 (x^2 - 1)\,dx = 0 \Rightarrow \left[\dfrac{x^3}{3} - x \right]_a^2 = 0$

$\left(\tfrac{8}{3} - 2 \right) - \left(\tfrac{a^3}{3} - a \right) = 0$

$\dfrac{8-6}{3} - \dfrac{a^3}{3} + a = 0$

$\dfrac{2}{3} - \dfrac{a^3}{3} + a = 0$

$\Rightarrow a - \dfrac{a^3}{3} = -\dfrac{2}{3}$

$3a - a^3 = -2$

$a(3 - a^2) = -2$

$a(a^2 - 3) = 2$

$a = -1$

> By trial and error since $a < 2$ try 1, 0, –1 and find that
> $-1((-1)^2 - 3)$
> $= -1(1 - 3)$
> $= -1(-2) = 2$

Alternative Method:

$$3a - a^3 = -2$$

By synthetic division $\Rightarrow 3a - a^3 + 2 = 0$

$$\Rightarrow a^3 - 3a - 2 = 0$$

$a^3 - 0a^2 - 3a - 2$

	1	0	−3	−2
−1		−1	1	2
	1	−1	−2	0

factors $(x + 1)(x^2 - x - 2)$

$= (x + 1)(x + 1)(x - 2)$

$x = -1$ or $x = 2$

hence $a = -1$

14. $f(x) = 0 \Rightarrow x(x^2 - 3)(x^2 + 4)(x^2 - 1) = 0$

$x = 0, \quad x = \pm\sqrt{3}, \quad x = \pm 1,$

Note: $x^2 + 4 = 0 \Rightarrow x^2 = -4$

no real roots $x \notin R$

s.s $\{-\sqrt{3}, -1, 0, \sqrt{3}, 1\}$

15. P(−2, 5), Q(2, −1), R(4, 2)

$m_{PR} = \dfrac{5 - 2}{-2 - 4} = \dfrac{3}{-6} = -\dfrac{1}{2}$

$m_{PR} \cdot m_{AQ} = -1$ if AQ ⊥ PR

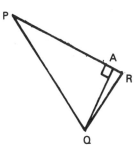

$m_1 m_2 = -1 \quad m_1 = -\dfrac{1}{2}, m_2 = 2,$ through Q(2, −1)

$$y + 1 = 2(x - 2)$$
$$y + 1 = 2x - 4$$
$$y = 2x - 5 \text{ is equation of Altitude}$$

16.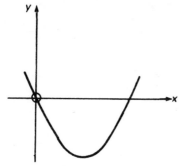

Graph of $6 + x - x^2$

(i) coefficient of x^2 negative $\Rightarrow$ shape

(ii) $ax^2 + bx + c$ cuts y at $(0, c)$ $c = 6$, point $(0, 6)$

2 main reasons why graph is not $f(x)$
$f(x) = ax^2 + bx + c$
   $a < 0$ parabola should be inverted
   c +ve $(0, c)$ should cut y at $(0, 6)$

17. General equation of a circle $x^2 + y^2 + 2gx + 2fy + c = 0$
$x^2 + y^2 + 12x - 4y + 15 = 0$
$2g = 12, -g = -6, \quad 2f = -4, -f = 2$
centre $(-g, -f) \quad r = \sqrt{g^2 + f^2 - c}$
centre $= (-6, 2) \quad r = \sqrt{6^2 + 2^2 - 15} = \sqrt{25} = 5$
Hence circle has centre $(-6, 2)$ radius 5.

18. *(a)* $f(x) = 2x^3 + mx$ (The gradient of the tangent $= 0$
    $f'(x) = 6x^2 + m = 0$ for s.v.   for stationary values)

$$f'\left(\frac{1}{\sqrt{2}}\right) = 6\left(\frac{1}{\sqrt{2}}\right)^2 + m = 0 \qquad \left(f'\left(-\frac{1}{\sqrt{2}}\right) \text{ is identical}\right)$$

$$6\left(\frac{1}{2}\right) + m = 0$$

$$3 + m = 0, m = -3$$

(b) Equation $f(x) = 2x^3 - 3x$
$f(-3) = 2(-3)^3 - 3(-3)$
$= -54 + 9 = -45$

19. *(a)* $A = PR^n$, $P = 600$, $R = \left(1 + \dfrac{6 \cdot 2}{100}\right)$, $n = 4$, $r = 6 \cdot 2\%$

        $A = 600(1 \cdot 062)^4 = £763.22$
        To double the investment $A = 1200$

(b) $600(R)^4 = 1200$

$R^4 = 2$

$R = \sqrt[4]{2}$

$R \fallingdotseq 1\cdot19$

$R = 1 + r\%$

$\Rightarrow r \fallingdotseq 19\%$

(Given a rate of 19%, A = £1203.20)

20. $x^3 - 3x^2 - x + a$, if divisible by $(x-3)$ then $f(3) = 0$

$$
\begin{array}{rrrr}
1 & -3 & -1 & a \\
 & 3 & 0 & -3 \\
\hline
1 & 0 & -1 & 0,
\end{array}
$$

$x = 3$ by synthetic division   (Remainder = 0)

$\Rightarrow -3 + a = 0$

$\Rightarrow a = 3$

$a = 3$

$f(x) = x^3 - 3x^2 - x + 3$

$f(x) = (x-3)(x^2 - 1)$

$\Rightarrow f(x) = (x-1)(x+1)(x-3)$ fully factorised

21. For function to have equal roots, discriminant = 0, discriminant = $b^2 - 4ac$

$3x^2 - 2x - c$ $b^2 - 4ac = 0$

$a = 3, b = -2, c = -c$

$b^2 - 4ac = 0 \Rightarrow (-2)^2 - 4(3)(-c) = 0$

$\Rightarrow 4 + 12c = 0$

$\Rightarrow \quad 12c = -4$

$c = \dfrac{-4}{12} = -\dfrac{1}{3}$

equation is

$$3x^2 - 2x + \frac{1}{3}$$

or $\dfrac{1}{3}(9x^2 - 6x + 1)$

$= \dfrac{1}{3}(3x - 1)^2$

22.

x	1	5
y	1·2	150

function of form $y = ax^n$

$\Rightarrow \log y = \log ax^n$

$\log y = \log a + \log x^n$

$\log y = \log a + n \log x$

$\Rightarrow \log y = n \log x + \log a$

$\log 150 = n \log (5) + \log a$

$\log 1\cdot2 = n \log (1) + \log a$

$\log 1 \quad = 0 \Rightarrow \log a = \log 1\cdot2$

$\Rightarrow a = 1\cdot2$

$\Rightarrow \log 150 = n \log (5) + \log 1\cdot2$

$2\cdot177 \quad = n(0\cdot699) + 0\cdot0792$

$2\cdot097 = (0\cdot699)n$

$\dfrac{2\cdot097}{0\cdot699} = n$

$n = 3, \quad a = 1\cdot2$

function of form $y = ax^n \Rightarrow y = 1\cdot2x^3$

check $150 = 1\cdot2(5)^3$

23.

$\overrightarrow{OP} = \underset{\sim}{p} \qquad \overrightarrow{OQ} = \underset{\sim}{q}$

P = (1, 4, −1), Q = (1, −2, 3)

$\underset{\sim}{p} = \begin{pmatrix} 1 \\ 4 \\ -1 \end{pmatrix} \quad \underset{\sim}{q} = \begin{pmatrix} 1 \\ -2 \\ 3 \end{pmatrix}$

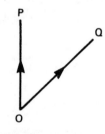

$p \cdot q = (1 \times 1) + (4 \times (-2)) + (-1 \times 3)$

$= 1 - 8 - 3 = -10$

$|\underset{\sim}{p}| = \sqrt{1^2 + 4^2 + (-1)^2}$

$= \sqrt{18}$

$|\underset{\sim}{q}| = \sqrt{1^2 + (-2)^2 + 3^2}$

$= \sqrt{14}$

$\dfrac{\overrightarrow{OP} \cdot \overrightarrow{OQ}}{|p||q|} = \cos P\hat{O}Q$

$= \dfrac{-10}{\sqrt{18}\sqrt{14}}$

$\cos$ −ve $\qquad P\hat{O}Q > 90°$ (obtuse)

$P\hat{O}Q = 180 - 50\cdot9 = 129\cdot1°$

OR

$P\hat{O}Q = 180 - 51 \quad = 129°$

24. *(a)* $U_{r+1} = mU_r + c,\ U_0 = 2,\ U_1 = -1,\ U_2 = 14$

$\quad U_1 \quad = m(U_0) + c \Rightarrow -1 = 2m + c \quad \text{①}$

$\quad U_2 \quad = m(U_1) + c \Rightarrow 14 = -m + c \quad \text{②}$

$\qquad\qquad \text{②} - \text{①}\ \ 15 = -3m \Rightarrow m = -5$

Substitute $m = -5$ in ① $\ -1 = 2(-5) + c$

$\qquad\qquad\qquad\qquad -1 = -10 + c$

$\qquad\qquad\qquad\qquad 9 = c$

$\quad m = -5,\ c = 9, \Rightarrow U_{r+1} = -5\,U_r + 9$

(b) $U_3 = -5U_2 + 9 \qquad\qquad$ To find U_{-1} use $U_0 = -5U_{-1} + 9,\ U_0 = 2$

$\quad\ \ = -5(14) + 9 \qquad\qquad\qquad 2 = -5U_{-1} + 9$

$\quad\ \ = -70 + 9 \qquad\qquad\qquad\ -7 = -5U_{-1}$

$\ U_3 = -61 \qquad\qquad\qquad\qquad\ \dfrac{7}{5} = U_{-1}$

(c) To find $U_r = U_{r+1} \Rightarrow\ U_r = -5U_r + 9$

$\qquad\qquad\qquad\qquad \Rightarrow 6U_r = 9$

$\qquad\qquad\qquad\qquad\quad U_r = \dfrac{9}{6},\ = \dfrac{3}{2}$

Check $-5\left(\dfrac{3}{2}\right) + 9 = -\dfrac{15}{2} + \dfrac{18}{2} = \dfrac{3}{2}$

$(U_r, U_{r+1}) = \left(\dfrac{3}{2}, \dfrac{3}{2}\right)$

25. $3 \sin x - 2 \cos x$ in form $k \sin (x - \alpha)$, (find $k \cos (x - \alpha)$

$\qquad\qquad$ and since $\sin (90 + x) = \cos x$, change to sin form later

$k = \sqrt{3^2 + (-2)^2} = \sqrt{13} \quad \tan \alpha = \dfrac{3}{-2} \begin{array}{l} \sin + \\ \cos -\text{ve} \end{array} \text{in ②} \quad \checkmark$

$\qquad\qquad\qquad\qquad \alpha = 123.7$

in $k \cos$ form $= \sqrt{13} \cos (x - 123.7)$

in $k \sin$ form $= \sqrt{13} \sin (90 + (x - 123.7))$

$\qquad\qquad\quad = \sqrt{13} \sin (x - 33.7)$

zeros at sin 0 and sin 180, $x - 33 \cdot 7 = 0$, $x = 33 \cdot 7$, $(33 \cdot 7, 0)$

$\qquad x - 33 \cdot 7 = 180$, $x = 213 \cdot 7$ $(213 \cdot 7, 0)$

maximum $\sqrt{13}$ at sin 90 $\Rightarrow x - 33 \cdot 7 = 90 \Rightarrow x = 123 \cdot 7$ $(123 \cdot 7, \sqrt{13})$

minimum $-\sqrt{13}$ at sin 270 $\Rightarrow x - 33 \cdot 7 = 270 \Rightarrow x = 303 \cdot 7$ $(303 \cdot 7, -\sqrt{13})$

when $x = 0$,

$\qquad \sqrt{13} \sin(x - 33 \cdot 7) = \sqrt{13} \sin(0 - 33 \cdot 7) = \sqrt{13} \times (-0 \cdot 555)$ $(0, -2)$

y intercept	x intercept	maximum	x intercept	minimum
Points $(0, -2)$	$(33 \cdot 7, 0)$	$(123 \cdot 7, \sqrt{13})$	$(213 \cdot 7, 0)$	$(303 \cdot 7, -\sqrt{13})$
				$(360, -2)$

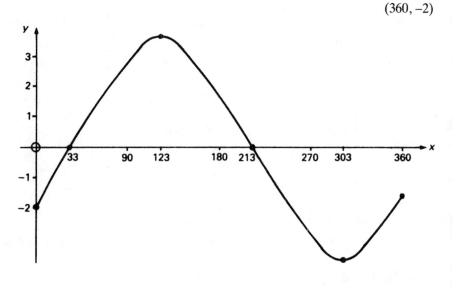

WORKED EXAMPLE — TEST PAPER E

1.

$$x^3 + 4x^2 + x \quad -t \qquad \text{If divisible by } x + 2 \text{ then } f(-2) = 0$$

-2	1	4	1	$-t$
		-2	-4	6
	1	2	-3	$6-t$

$\Rightarrow \quad 6 - t = 0, \Rightarrow t = 6$

$t = 6, \Rightarrow \quad$ | 1 | 2 | -3 | 0

$f(x) = (x + 2)(x^2 + 2x - 3)$
$f(x) = (x + 2)(x + 3)(x - 1)$ fully factorised

2. By Pythagoras' theorem

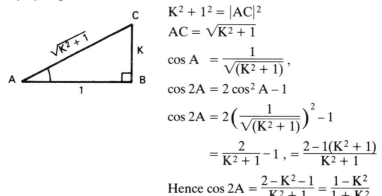

$K^2 + 1^2 = |AC|^2$

$AC = \sqrt{K^2 + 1}$

$\cos A \ = \dfrac{1}{\sqrt{(K^2 + 1)}},$

$\cos 2A = 2 \cos^2 A - 1$

$\cos 2A = 2\left(\dfrac{1}{\sqrt{(K^2 + 1)}}\right)^2 - 1$

$\qquad = \dfrac{2}{K^2 + 1} - 1, = \dfrac{2 - 1(K^2 + 1)}{K^2 + 1}$

Hence $\cos 2A = \dfrac{2 - K^2 - 1}{K^2 + 1} = \dfrac{1 - K^2}{1 + K^2}$

3. *(a)* General equation of a circle $x^2 + y^2 + 2gx + 2fy + c = 0$

$\qquad\qquad\qquad\qquad$ centre $(-g, -f)$, radius $= \sqrt{g^2 + f^2 - c}$

$x^2 + y^2 - 6x + 8y = 0 \qquad$ centre $(3, -4)$, $r = \sqrt{3^2 + (-4)^2} = 5$

$\qquad\qquad\qquad\qquad$ centre $(3, -4)$, radius $= 5$

(b)

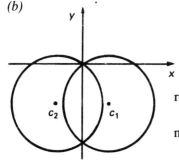

reflected in y-axis $c_1 \rightarrow c_2 \Rightarrow c_2 = (-3, -4)$

$\qquad\qquad\qquad\qquad$ radius unchanged

new equation $= x^2 + y^2 + 6x + 8y = 0$

4. $\frac{2+x}{2} - (2-x) < 5$

$$2 + x - 2(2-x) < 10$$
$$2 + x - 4 + 2x < 10$$
$$3x - 2 < 10$$
$$3x < 12$$
$$x < 4$$

5. $f(x) = 3 \sin 2x$

$f'(x) = 6 \cos 2x,$

$$x = \frac{\pi}{6} \Rightarrow 6 \cos 2\left(\frac{\pi}{6}\right)$$

$$= 6 \cos \left(\frac{\pi}{3}\right) = 6 \times \frac{1}{2} = 3$$

$f'\left(\frac{\pi}{6}\right) = 3$

6. $3^x = 100$ can be solved by trial and error

Log method:

$$\log 3^x = \log 100$$
$$x \log 3 = \log 100$$
$$x = \frac{\log 100}{\log 3} \Rightarrow x \doteqdot 4{\cdot}19$$

$$\text{Test } 3^{4{\cdot}19} = 99{\cdot}8$$
$$3^{4{\cdot}2} = 100{\cdot}90$$
$$x \doteqdot 4{\cdot}19 \text{ to 2 decimal places}$$

7. By Pythagoras' theorem $\quad \sqrt{2^2 + 1^2} = \text{hypotenuse} = \sqrt{5} \text{ hypotenuse.}$

Hence $\cos A = \dfrac{2}{\sqrt{5}} = \dfrac{2\sqrt{5}}{5}$

$\qquad\qquad = p\sqrt{5} \text{ where } p = \dfrac{2}{5}$

8. $H = 3\cos\left(2x - \frac{\pi}{3}\right)$ $\quad 0 \leqslant x \leqslant 2\pi$

Maximum value $= 3$

$$\text{when } \left(2x - \frac{\pi}{3}\right) = 0, \overset{2\pi}{360}, \overset{4\pi}{720}$$

$2x - \frac{\pi}{3} = 0$

$\quad 2x = \frac{\pi}{3}, x = \frac{\pi}{6}$

$2x - \frac{\pi}{3} = 2\pi$

$\quad 2x = 2\pi + \frac{\pi}{3} = \frac{7\pi}{3} \quad x = \frac{7\pi}{6}$

$2x - \frac{\pi}{3} = 4\pi$

$\quad 2x = 4\pi + \frac{\pi}{3} = \frac{13\pi}{3}, x = \frac{13\pi}{6}$

$$\frac{13\pi}{6} > 2\pi$$

minimum $= -3$ when $2x - \frac{\pi}{3} = \pi, 3\pi$

$2x = \frac{\pi}{3} + \pi,$ $\qquad\qquad 2x = \frac{\pi}{3} + 3\pi$

$2x = \frac{4\pi}{3}, x = \frac{4\pi}{6}$ $\qquad\qquad 2x = \frac{10\pi}{3}, x = \frac{10\pi}{6}$

$\quad x = \frac{2\pi}{3}$ $\qquad\qquad\qquad x = \frac{5\pi}{3}$

maximum $= 3$, when $x = \frac{7\pi}{6}$ or $\frac{\pi}{6}$

minimum value is -3 at $\frac{2\pi}{3}$ and $\frac{5\pi}{3}$

Hence coordinates $(x, y) = \left(\frac{\pi}{6}, 3\right), \left(\frac{2\pi}{3}, -3\right), \left(\frac{7\pi}{6}, 3\right), \left(\frac{5\pi}{3}, -3\right)$

9.

$$f(x) = (2x + \sqrt{x})^3$$

$$f(x) = (2x + x^{1/2})^3 \quad \text{(By chain rule method.)}$$

$$f'(x) = 3(2x + x^{1/2})^2\left(2 + \tfrac{1}{2}x^{-1/2}\right)$$

$$= 3\left(2 + \frac{1}{2\sqrt{x}}\right)\left(2x + \sqrt{x}\right)^2$$

hence $f'(x) = 3\left(2 + \dfrac{1}{2\sqrt{x}}\right)\left(2x + \sqrt{x}\right)^2$

and $\quad f'(4) = 3\left(2 + \dfrac{1}{2\sqrt{4}}\right)\left(2(4) + \sqrt{4}\right)^2$

$$= 3(2\tfrac{1}{4})(10)^2$$

$$= 3(225)$$

$$= 675$$

10. $f(x) = x(x + 2)(x^2 - 3)(x^2 + 1)(x^2 - 4) = 0,$

Note: $x^2 + 1 = 0, \Rightarrow x^2 = -1,$ (not real) $x \notin R$

$\Rightarrow f(x) = 0, x = 0, x = -2, x = \pm\sqrt{3}, x = \pm 2$

Hence S.S. $\{-\sqrt{3}, -2, 0, \sqrt{3}, 2\}$

11. $h(x) = g(f(x)) \quad g(x) = -x^2 + x + 2 \quad f(x) = 2x - 1$

$$g(f(x)) = g((2x - 1)) = -(2x - 1)^2 + (2x - 1) + 2$$

$$= -(4x^2 - 4x + 1) + 2x - 1 + 2$$

$$= -4x^2 + 4x - 1 + 2x + 1$$

$$= -4x^2 + 6x$$

12.

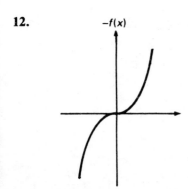

$-f(x)$

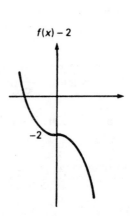

$f(x) - 2$

-2

13. $\underset{\sim}{a} = \begin{pmatrix} 3 \\ 1 \\ 3 \end{pmatrix}$ $\underset{\sim}{b} = \begin{pmatrix} -2 \\ 2 \\ -2 \end{pmatrix}$ $P = \frac{1}{5} (3\underset{\sim}{a} + 2\underset{\sim}{b})$

$$= \frac{1}{5} \left[\begin{pmatrix} 9 \\ 3 \\ 9 \end{pmatrix} + \begin{pmatrix} -4 \\ 4 \\ -4 \end{pmatrix} \right] = \frac{1}{5} \begin{pmatrix} 5 \\ 7 \\ 5 \end{pmatrix}$$

$$P = \left(1, \frac{7}{5}, 1 \right)$$

14. P(2, 7), Q(0, −1), R(−5, 4)

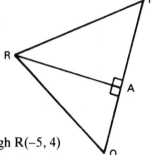

$$m_{PQ} = \frac{7 - (-1)}{2 - 0} = \frac{8}{2} = 4$$

$m_{PQ} \cdot m_{AR} = -1 \Rightarrow m_{AR} = -\frac{1}{4}$ through R(−5, 4)

$$y - 4 = -\frac{1}{4}(x + 5)$$

$$4y - 16 = -x - 5$$

$$4y + x = 11 \text{ equation of altitude AR}$$

15. *(a)* $6\sin^2 x - \sin x - 2$

> let $\sin x = x$,
> $6x^2 - x - 2$

$(3\sin x - 2)(2\sin x + 1) = 0$ $(3x - 2)(2x + 1) = 0$

(b) $(3\sin x - 2)(2\sin x + 1) = 0 \Rightarrow \sin x = \frac{2}{3}$ or $\sin x = -\frac{1}{2}$

$\sin x$ positive $\sin x$ negative
in quadrants ①, ②   in quadrants ③, ④

S.S. {41·8, 138·2, 210, 330}

16. Let P, Q, R represent the given points P(1, −1), Q(a, 2), R(b, 1)

$$\underset{\sim}{p} = \begin{pmatrix} 1 \\ -1 \end{pmatrix}, \; \underset{\sim}{q} = \begin{pmatrix} a \\ 2 \end{pmatrix}, \; \underset{\sim}{r} = \begin{pmatrix} b \\ 1 \end{pmatrix}$$

$$\vec{PQ} = \underset{\sim}{q} - \underset{\sim}{p} \qquad\qquad \vec{QR} = \underset{\sim}{r} - \underset{\sim}{q}$$

$$= \begin{pmatrix} a \\ 2 \end{pmatrix} - \begin{pmatrix} 1 \\ -1 \end{pmatrix} \qquad\qquad = \begin{pmatrix} b \\ 1 \end{pmatrix} - \begin{pmatrix} a \\ 2 \end{pmatrix}$$

$$= \begin{pmatrix} a-1 \\ 3 \end{pmatrix} \qquad\qquad\qquad = \begin{pmatrix} b-a \\ -1 \end{pmatrix}$$

If collinear then vectors are parallel.

hence $\begin{pmatrix} a-1 \\ 3 \end{pmatrix} = k\begin{pmatrix} b-a \\ -1 \end{pmatrix}$

where $k = -3$

hence $-3\begin{pmatrix} b-a \\ -1 \end{pmatrix} = \begin{pmatrix} a-1 \\ 3 \end{pmatrix}$

$$\Rightarrow -3b + 3a = a - 1$$
$$\Rightarrow -3b + 2a = -1$$
$$\text{or} \quad 3b - 2a = 1$$

Alternative Method:

P(1, −1), Q(a, 2), R(b, 1) Q is a shared point

$$m_{PQ} = \frac{3}{a-1}, \quad m_{QR} = \frac{1}{a-b}$$

$$m_{PQ} = m_{QR} \Rightarrow 3(a-b) = 1(a-1)$$
$$3a - 3b = a - 1$$
$$2a = 3b - 1$$
$$1 = 3b - 2a$$
$$3b - 2a = 1$$
$$\Rightarrow \text{if collinear } 3b - 2a = 1$$

17. $4x^3 + mx = f(x) \qquad f'(x) = 0$ for stationary value(s)

$$f'(x) = 12x^2 + m = 0$$
$$12x^2 = -m$$
$$x^2 = \frac{-m}{12}$$

$$x = \pm\frac{3}{2}, \; \left(-\frac{3}{2}\right)^2 = \left(\frac{3}{2}\right)^2$$

$$x = \frac{3}{2}, \quad \frac{9}{4} = \frac{-m}{12} \quad \Rightarrow \quad \frac{108}{4} = -m, \quad 27 = -m, \quad m = -27$$

hence equation is $4x^3 - 27x$

18. $(2\sqrt{3} + 3\sqrt{2})^2 = (2\sqrt{3} + 3\sqrt{2})(2\sqrt{3} + 3\sqrt{2})$
$$= 2\sqrt{3}(2\sqrt{3} + 3\sqrt{2}) + 3\sqrt{2}(2\sqrt{3} + 3\sqrt{2})$$
$$12 + 6\sqrt{6} + 6\sqrt{6} + 18$$
$$= 30 + 12\sqrt{6}$$

19. (a) $A = PR^n$, $P = 1000$, $R = 1 + \frac{5 \cdot 8}{100} = 1 \cdot 058$, $n = 7$

$A = 1000(1 \cdot 058)^7 = £1483.88$

(b) Half yearly $n = 14$, $R = 1 + \frac{2 \cdot 9}{100} \Rightarrow A = 1000(1 \cdot 029)^{14} = £1492.16$,

difference $= £8.28$

20. $\int_{-3}^{0} (\text{upper} - \text{lower})\, dx + \int_{0}^{5} (\text{upper} - \text{lower})\, dx$

$= \int_{-3}^{0} (f_2(x) - f_1(x))\, dx + \int_{0}^{5} (f_1(x) - f_2(x))\, dx$

21. (a) $U_{r+1} = K U_r + t \qquad U_0 = 0,\ U_1 = 2,\ U_2 = -4$

$U_1 \quad = K U_0 + t \Rightarrow 2 = t$

$U_2 \quad = K U_1 + t \Rightarrow -4 = K(2) + t,\ t = 2$

$\qquad\qquad\qquad\qquad -4 = 2K + 2$

$\qquad\qquad\qquad\qquad -6 = 2K$

$\qquad\qquad t = 2, \quad K = -3$

$U_{r+1} = K U_r + t \Rightarrow U_{r+1} = -3 U_r + 2$

(b) $\qquad U_4 = -3 U_3 + 2$

$\qquad\quad U_3 = -3 U_2 + 2$

$\qquad\qquad\ = -3(-4) + 2$

$\qquad\quad U_3 = 14$

hence $U_4 = -3(14) + 2$

$\qquad\qquad = -42 + 2$

$\qquad\qquad = -40$

To find U_{-1} use $U_0 = -3 U_{-1} + 2$

$\qquad\qquad\qquad 0 = -3 U_{-1} + 2$

$\qquad\qquad\quad -2 = -3 U_{-1}$

$\qquad\qquad\quad \frac{2}{3} = U_{-1}$

$\qquad\qquad\quad U_4 = -40$

$\qquad\qquad\quad U_{-1} = \frac{2}{3}$

22. $4x^2 + 4x + 5$

$4\left(x^2 + x + \frac{5}{4}\right) = 4\left(x + \frac{1}{2}\right)^2 + 4\left(-\frac{1}{4} + \frac{5}{4}\right)$

$\qquad\qquad\qquad\quad = 4\left(x + \frac{1}{2}\right)^2 + 4$

minimum value $= 4$ when $x = -\frac{1}{2}$

23. $y = 3x^2 - 2x + 1$

$$\frac{dy}{dx} = 6x - 2 = m \tan$$

parallel to $y = x - 3$
$$\Rightarrow m = 1$$
$$\Rightarrow 6x - 2 = 1$$
$$6x = 3$$
$$x = \frac{1}{2}, \; f\left(\frac{1}{2}\right) = 3\left(\frac{1}{2}\right)^2 - 2\left(\frac{1}{2}\right) + 1$$
$$\frac{3}{4} - 1 + 1 = \frac{3}{4}$$

Point $\left(\frac{1}{2}, \frac{3}{4}\right)$ $m = 1$

$$y - \frac{3}{4} = 1\left(x - \frac{1}{2}\right)$$
$$4y - 3 = 4\left(x - \frac{1}{2}\right)$$
$$4y - 3 = 4x - 2$$
$$4y - 4x = 1 \quad \text{or} \quad 4y = 4x + 1 \quad \text{(equation of the tangent)}$$

24. $-4 \sin x - 3 \cos x$ in form $k \cos(x - \lambda)$

$k = \sqrt{(-4)^2 + (-3)^2} = 5$

$\tan \lambda = \dfrac{-4}{-3} \quad \begin{matrix} \sin -ve \\ \cos -ve \end{matrix}$ in quadrant 3

$\lambda = 233°$

in form $k \cos(x - \lambda)$ gives $5 \cos(x - 233°)$

Points

when $x = 0$, $5 \cos(x - 233) = 5 \cos(127) = -3$ $(0, -3)$

when $k \cos(x - 233) = 0$, $x - 233 = 90 \Rightarrow x = 323$, $(323, 0)$

or $x - 233 = 270 \Rightarrow x = 503 - 360 = 143$ $(143, 0)$

maximum = 5 when $x - 233 = 0 \Rightarrow x = 233$ $(233, 5)$

minimum = -5 when $x - 233 = 180 \Rightarrow x = 413 - 360 = 53$ $(53, -5)$

y intercepts x intercepts

$(0, -3)$ $(143, 0)(323, 0)$

minimum   maximum

$(53, -5)$ $(233, 5)$

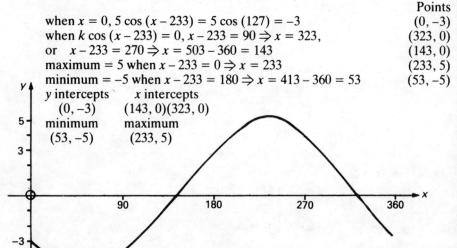

82

25. *(a)* $f(x) = 2x^2 - bx + 3$ for equal roots $b^2 - 4ac = 0$

$\qquad a = 2, b = -b, c = 3$

$\qquad\qquad b^2 - 4ac$

$\qquad = (-b)^2 - 4(2)(3)$

$\qquad\qquad b^2 - 24 = 0$

$\qquad\qquad\qquad b^2 = 24$

$\qquad\qquad\qquad b = \pm\sqrt{24}$

$\qquad\qquad\qquad b = 2\sqrt{6}$ or $-2\sqrt{6}$

(b) for real roots $b^2 - 4ac \geqslant 0$, $\qquad 4ac = 24$

$\qquad b \leqslant -2\sqrt{6}$ or $b \geqslant 2\sqrt{6}$

since $(-b)^2 > 24$ for these values

then $b^2 - 4ac > 0$

(c) for no real roots

$\qquad\qquad b^2 - 4ac < 0$

$\Rightarrow \quad b^2 - 24 \ < 0$

$\qquad\quad b^2 \qquad < 24$

$\qquad -2\sqrt{6} < b < 2\sqrt{6}$

(d) $f(x)$ cuts y-axis at $(0, 3)$

---

equal roots $b = \pm 2\sqrt{6}$

$\dfrac{-b \pm \sqrt{0}}{2a} = \dfrac{-b}{2a} = \dfrac{-2\sqrt{6}}{4} = \dfrac{-\sqrt{6}}{2}$ or $\dfrac{\sqrt{6}}{2}$

---

equal roots $\qquad\qquad$ real roots $\qquad\qquad$ no real roots

83

WORKED EXAMPLE — TEST PAPER F

1. $f(x) = x^3 - 3x - 5$

$f'(x) = 3x^2 - 3 = m \tan = 0$ for S.V.

$\quad\quad 3(x^2 - 1) = 0$

$\quad\quad\quad\quad x = \pm 1$

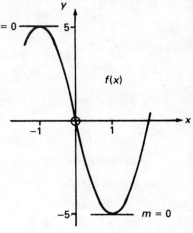

x	$\rightarrow$	-1	$\rightarrow$	1	$\rightarrow$
$(x^2 - 1)$	$+$	0	$-$	0	$+$
$f'(x)$	$\nearrow$	$\rightarrow$	$\searrow$	$\rightarrow$	$\nearrow$

shape

$f'(x)$ decreasing $-1 < x < 1$

Alternative Method:

Using second derivative

$f''(x) = 6x$

$f''(-1)$ −ve $\quad$ max T.P. $f(-1)$

$f''(1)$ +ve $\quad$ min T.P. $f(1)$

$f(x)$ decreasing for $\{-1 < x < 1\}$

2. $3xy = -2$

$y = \dfrac{-2}{3x}$ $\quad x = 1, y = -\dfrac{2}{3}, x = -1, y = \dfrac{2}{3}$

$\quad\quad P\left(1, -\dfrac{2}{3}\right), Q\left(-1, \dfrac{2}{3}\right)$

$\quad\quad\quad m_{PQ} = \dfrac{\frac{4}{3}}{-2} = -\dfrac{4}{6} = -\dfrac{2}{3}$

$\quad\quad$ gradient of PQ $= -\dfrac{2}{3}$

3. $\quad \cos 3x + \cos x$

$= \cos(2x + x) + \cos(2x - x)$

$= \cos 2x \cos x - \sin 2x \sin x$

$\quad + \cos 2x \cos x + \sin 2x \sin x$

$= 2 \cos 2x \cos x$

$\quad\quad$ Hence $\cos 3x + \cos x = 2 \cos 2x \cos x$

84

4. $2x^2 + x + 2$

$= 2\left(x^2 + \frac{1}{2}x + 1\right)$

$= 2\left(x + \frac{1}{4}\right)^2 + 2\left(-\frac{1}{16} + 1\right)$

$= 2\left(x + \frac{1}{4}\right)^2 + 2\left(\frac{15}{16}\right)$

$= 2\left(x + \frac{1}{4}\right)^2 + \frac{15}{8}$

minimum value $= \frac{15}{8}$ when $x = -\frac{1}{4}$, $\left(-\frac{1}{4}, \frac{15}{8}\right)$ min, T.P.

5. $\frac{x - 2y}{3} = \frac{y - 2x}{2} \Rightarrow 2(x - 2y) = 3(y - 2x)$

$$2x - 4y = 3y - 6x$$

$$8x = 7y$$

$$x = \frac{7}{8}y$$

To find value of $\frac{7x - 2y}{3x + y}$

Substitute $x = \frac{7}{8}y$

$$= \frac{7\left(\frac{7}{8}\right)y - 2y}{3\left(\frac{7}{8}\right)y + y}$$

$$= \frac{\left(\frac{49}{8} - 2\right)y}{\left(\frac{21}{8} + 1\right)y} = \frac{\frac{49 - 16}{8}}{\frac{21 + 8}{8}} = \frac{33}{29}$$

Hence $\frac{7x - 2y}{3x + y} = \frac{33}{29}$

6. $f(x) = x^2 - 3$, $g(x) = 2 - x$

$f(g(x)) = f(2-x) = (2-x)^2 - 3$
$$= 4 - 4x + x^2 - 3$$

$f(g(x)) = x^2 - 4x + 1$

$f(g(2)) = 2^2 - 4(2) + 1$
$$= 4 - 8 + 1 = -3$$

Alternative Method:
$g(2) = 2 - 2 = 0$
$f(0) = 0^2 - 3 = -3$

7.

x	< 2	2	> 2
$f'(x)$	$+$	0	$-ve$
plot	above	on	below

in relation to x-axis

intercepts x-axis at $(2, 0)$

⌢ shape $-x^2$ $f'(x) \Rightarrow m$ is $-ve$ ↘

$f(x)$ quadratic, $f'(x)$ linear

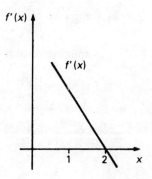

8. $\int_1^4 \left(\sqrt{x} + \dfrac{1}{2\sqrt{x}} \right) dx = \int_1^4 \left(x^{1/2} + \dfrac{1}{2}(x)^{-1/2} \right) dx$

$$= \left[\dfrac{2x^{3/2}}{3} + x^{1/2} \right]_1^4$$

$$= \left(\dfrac{16}{3} + 2 \right) - \left(\dfrac{2}{3} + 1 \right)$$

$$= \dfrac{14}{3} + 1 = \dfrac{17}{3} \text{ or } 5\dfrac{2}{3}$$

9. $f(x) = 2\cos 3x$

$f'(x) = -6\sin 3x$, $f'\left(\dfrac{\pi}{2}\right) = -6\sin\left(\dfrac{3\pi}{2}\right)$

$$= -6 \times -1 = 6$$

10. General equation of a circle $x^2 + y^2 + 2gx + 2fy + c = 0$

centre $(-g, -f)$

$x^2 + y^2 - 6x + 8y = 0$, $2g = -6 \Rightarrow -g = 3$

$\qquad\qquad\qquad\qquad 2f = 8 \Rightarrow -f = -4$

$(-g, -f) \Rightarrow$ centre $= (3, -4)$

$\qquad\qquad C_1 \rightarrow C_2$ under reflection in x-axis

$\qquad\qquad\qquad \Rightarrow (3, -4) \rightarrow (3, 4)$

$C_2 = (3, 4)$, $\qquad -g = 3 \Rightarrow 2g = -6$

$\qquad\qquad\qquad -f = 4 \Rightarrow 2f = -8$

equation of circle under reflection in x-axis $= x^2 + y^2 - 6x - 8y = 0$

11. $x^2 + 100 = x^2 - 100 + 200$

$$\frac{x^2 + 100}{x + 10} = \frac{(x + 10)(x - 10) + 200}{x + 10} = x - 10 + \frac{200}{x + 10}$$

$$k = 200$$

12. If points are collinear then lines are parallel.

Hence $\overrightarrow{AB} = k\,\overrightarrow{BC}$

$A(3, -1)$, $B(a, 2)$, $C(b, 5)$

$$\underset{\sim}{a} = \begin{pmatrix} 3 \\ -1 \end{pmatrix}, \underset{\sim}{b} = \begin{pmatrix} a \\ 2 \end{pmatrix}, \underset{\sim}{c} = \begin{pmatrix} b \\ 5 \end{pmatrix}$$

$$\overrightarrow{AB} = \underset{\sim}{b} - \underset{\sim}{a} = \begin{pmatrix} a - 3 \\ 3 \end{pmatrix}, \overrightarrow{BC} = \begin{pmatrix} b - a \\ 3 \end{pmatrix}$$

$\qquad \Rightarrow a - 3 = b - a$

$\qquad \Rightarrow 2a = b + 3$

$\qquad \Rightarrow 2a - b = 3$

Alternative Method:

$A(3, -1)$, $B(a, 2)$, $C(b, 5)$

If points are collinear then gradient of AB = gradient of BC.

$$m_{AB} = \frac{3}{a - 3}, m_{BC} = \frac{3}{b - a}$$

$\qquad \Rightarrow \dfrac{3}{a - 3} = \dfrac{3}{b - a}$

$\qquad \Rightarrow a - 3 = b - a$

$\qquad \Rightarrow 2a = 3 + b$

$\qquad \Rightarrow 2a - b = 3$

13. $K = 3 \cos \left(3x - \dfrac{\pi}{2} \right)$ $0 < x < 2\pi$

maximum value $= 3$ when $\cos \left(3x - \dfrac{\pi}{2} \right) = 1$

$\cos 0 = 1, \cos 2\pi = 1, \cos 4\pi = 1$

$\Rightarrow \left(3x - \dfrac{\pi}{2} \right) = 0, 2\pi \text{ or } 4\pi$

$\Rightarrow 3x = 0 + \dfrac{\pi}{2}, 2\pi + \dfrac{\pi}{2}, \text{ or } 4\pi + \dfrac{\pi}{2}$

$\Rightarrow 3x = \dfrac{\pi}{2}, \dfrac{5\pi}{2} \text{ or } \dfrac{9\pi}{2}$

$\Rightarrow \quad x = \dfrac{\pi}{6}, \dfrac{5\pi}{6} \text{ or } \dfrac{3\pi}{2}$

minimum value $= -3$ when $\cos \left(3x - \dfrac{\pi}{2} \right) = -1$

$\cos \pi = -1, \cos 3\pi = -1, \cos 5\pi = -1$

$\overset{\Rightarrow}{} \left(3x - \dfrac{\pi}{2} \right) = \pi, 3\pi, 5\pi$

$\Rightarrow 3x = \pi + \dfrac{\pi}{2}, 3\pi + \dfrac{\pi}{2}, 5\pi + \dfrac{\pi}{2}$

$\Rightarrow 3x = \dfrac{3\pi}{2}, \dfrac{7\pi}{2}, \dfrac{11\pi}{2}$

$\Rightarrow \quad x = \dfrac{\pi}{2}, \dfrac{7\pi}{6}, \dfrac{11\pi}{6}$

solution set $\left\{ \dfrac{\pi}{6}, \dfrac{\pi}{2}, \dfrac{5\pi}{6}, \dfrac{7\pi}{6}, \dfrac{3\pi}{2}, \dfrac{11\pi}{6} \right\}$

or coordinates are $\left(\dfrac{\pi}{6}, +3 \right), \left(\dfrac{\pi}{2}, -3 \right), \left(\dfrac{5\pi}{6}, +3 \right)$

$\left(\dfrac{7\pi}{6}, -3 \right), \left(\dfrac{3\pi}{2}, +3 \right), \left(\dfrac{11\pi}{6}, -3 \right)$

14. *(a)* $A = PR^n, P = 800, R = 1 + \dfrac{7 \cdot 8}{100}, n = 5, r = 7 \cdot 8\%$

$A = 800(1 \cdot 078)^5$

$A = £1164 \cdot 62$

(b). To double the investment

amount $= 2 \times 800 = 1600$

$A = PR^n$, $P = 800$, $R = 1 \cdot 078$, $A = 1600$

$800(1 \cdot 078)^n = 1600$

$(1 \cdot 078)^n = 2$

[This can be solved by trial and error on the calculator.]

Log method

$(1 \cdot 078)^n \quad = 2$

$\log 1 \cdot 078^n = \log 2$

$n \log 1 \cdot 078 = \log 2$

$$n = \frac{\log 2}{\log 1 \cdot 078}$$

$$n \doteqdot 9 \cdot 23$$

Hence it would take 9 years to almost double the investment.

$$800(1 \cdot 078)^9 = £1572.75$$

$$\text{or } 10 \text{ years to exceed } £1600$$

15. $A(-2, 5)$, $B(2, -1)$, $C(6, 5)$. Let centroid $= M$

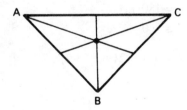

mid pt. $BC = \left(\dfrac{2+6}{2}, \dfrac{-1+5}{2} \right)$

$= (4, 2)$ and call this D

$AM:MD = 2:1$

$$M = \frac{1}{3}\left(\binom{8}{4} + \binom{-2}{5} \right), = \frac{1}{3}\binom{6}{9}$$

$$= \binom{2}{3} = \underset{\sim}{m}$$

$\Rightarrow \qquad M = (2, 3)$

$\Rightarrow$ coordinates of centroid $= (2, 3)$

Alternative Method:

$$\overrightarrow{OM} = \frac{1}{3}(\underset{\sim}{a} + \underset{\sim}{b} + \underset{\sim}{c})$$

$$= \frac{1}{3}\left(\begin{pmatrix} -2 \\ 5 \end{pmatrix} + \begin{pmatrix} 2 \\ -1 \end{pmatrix} + \begin{pmatrix} 6 \\ 5 \end{pmatrix}\right)$$

$$= \frac{1}{3}\begin{pmatrix} 6 \\ 9 \end{pmatrix} \Rightarrow M = (2, 3)$$

Coordinates of centroid = (2, 3)

---

16. $3\sin^2 x - \sin x - 2 = 0,$ $\qquad\qquad$ Let $\sin x = x$

$$3x^2 - x - 2 = 0$$
$$(3x + 2)(x - 1) = 0$$

$3\sin^2 x - \sin x - 2 = 0$

$\Rightarrow (3\sin x + 2)(\sin x - 1) = 0$

$\Rightarrow \sin x = -\frac{2}{3}$ $\qquad$ or $\sin x = 1$

$\qquad\qquad\qquad\qquad\qquad x = 90$

$$\sin^{-1}\frac{2}{3} = 41{\cdot}8°$$

$$x = 180 + 41{\cdot}8$$

$$x = 360 - 41{\cdot}8$$

$\sin$ –ve in ③ ④ $\qquad\qquad\qquad$ S.S. $\{90, 221{\cdot}8, 318{\cdot}2\}$

---

17. $e^x = 35$ $\quad$ take $e \approx 2{\cdot}72$

Log method.

$\log e^x = \log 35$

$x \log e = \log 35$

$$x = \frac{\log 35}{\log e} \approx 3{\cdot}56 \quad x \approx 3{\cdot}56 \text{ to 2 decimal places.}$$

* $e^x = 35 \Rightarrow 2{\cdot}72^x = 35$

This can be found by method of trial and error on calculator.

18. $f(x) = 0 \Rightarrow x(x^2 + 4)(x^2 - 3)(x^2 - 1) = 0$

$$\Rightarrow x = 0, x = \pm\sqrt{3}, x = \pm\sqrt{1}$$

$$\text{S.S. } \{-\sqrt{3}, -1, 0, 1, \sqrt{3}\}$$

Note $x^2 + 4 = 0 \Rightarrow x^2 = -4$

No real solution, $x \notin R$

19. $f(x) = x^3 + 3x^2 - 4x + q$

If $(x - 2)$ is a factor of $f(x)$

then $f(2) = 0$

By synthetic division

x^3	$+$	$3x^2$	$-$	$4x$	$+$	q
1		3		-4		q
2		2		10		12
1		5		6		$12 + q$

$\Rightarrow 12 + q = 0,$

$q = -12$

$f(x) = x^3 + 3x^2 - 4x - 12$

20. P(-1, 5), Q(-3, 2), R(9, -1)

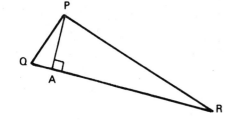

if AP ⊥ to QR

$m_{AP} \cdot m_{QR} = -1$

$$m_{QR} = \frac{2 - (-1)}{-3 - 9} = \frac{3}{-12} = \frac{-1}{4}$$

$m_{AP} = 4$ through P(-1, 5)

$$y - 5 = 4(x + 1)$$

$$y - 5 = 4x + 4 \Rightarrow y = 4x + 9 \text{ equation of altitude AP.}$$

21. *(a)* $U_{r+1} = KU_r + t$ $U_0 = 2, U_1 = -2, U_2 = 10$

$$U_1 = KU_0 + t \Rightarrow -2 = (2)K + t \quad ①$$

$$U_2 = KU_1 + t \Rightarrow 10 = (-2)K + t \quad ②$$

$$\text{add} \Rightarrow 8 = 2t, \quad t = 4$$

Substitute $t = 4$ in ① $-2 = 2K + t \Rightarrow -2 = 2K + 4$

$$\Rightarrow 2K = -6$$

$$K = -3$$

$U_{r+1} = KU_r + t, K = -3, t = 4$

$U_{r+1} = -3U_r + 4$

(b) To find $U_{r+1} = U_r \Rightarrow -3U_r + 4 = U_r$

$$\Rightarrow 4U_r = 4$$

$$\Rightarrow U_r = 1$$

Test: $U_r = 1, U_{r+1} = -3(1) + 4 = 1$

22. $V = 1125 \text{ cm}^3$

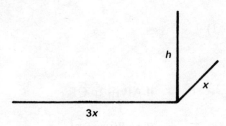

Let $l = 3x, b = x, h = h$

$V = lbh = 3x(x)h$

$$V = 3x^2h \Rightarrow h = \frac{V}{3x^2}$$

$V = 1125 \Rightarrow h = \dfrac{1125}{3x^2}$

$$\text{Surface area} = 2(3x^2) + 2(xh) + 2(3xh)$$

$$= 6x^2 + 8xh$$

$$= 6x^2 + 8x\left(\frac{1125}{3x^2}\right) \quad \left(\text{since } h = \frac{1125}{3x^2}\right)$$

$$A(x) = 6x^2 + \frac{9000}{3x}$$

$$= 6x^2 + 3000x^{-1}$$

Check $f''(x)$ $\qquad A'(x) = 12x - 3000x^{-2}$

$$= 12 + \frac{6000}{x^3} \; \text{+ve} \curvearrowright \qquad \Rightarrow 12x - \frac{3000}{x^2} = m \tan = 0 \text{ for S.V.}$$

hence a minimum $f(6{\cdot}3)$

$$\Rightarrow 12x = \frac{3000}{x^2} \Rightarrow 12x^3 = 3000$$

$$x^3 = 250$$

$$x = \sqrt[3]{250}$$

$$x = 6{\cdot}299$$

$$x \doteqdot 6{\cdot}3$$

Hence L = 18·90, B = 6·30, H = 9·45

Each rounded to 2 decimal places

Check $L \times B \times H = V$

$$18{\cdot}9 \times 6{\cdot}3 \times 9{\cdot}45 = 1125{\cdot}2$$

23. $f(x) = 3\cos 2x$ $\quad$ max $= 3$ when $2x = 0, 2\pi, 4\pi$

$$x = 0, \pi, 2\pi$$

$$\text{min} = -3 \text{ when } 2x = \pi, 3\pi$$

$$x = \frac{\pi}{2}, \frac{3\pi}{2}$$

x intercept, $\cos 2x = 0$, $2x = \dfrac{n\pi}{2}$ (for n odd)

$$2x = \frac{\pi}{2}, \frac{3\pi}{2}, \frac{5\pi}{2}, \frac{7\pi}{2}$$

$$x = \frac{\pi}{4}, \frac{3\pi}{4}, \frac{5\pi}{4}, \frac{7\pi}{4}$$

Points to plot:

$(0, 3), \left(\frac{\pi}{4}, 0\right), \left(\frac{\pi}{2}, -3\right), \left(\frac{3\pi}{4}, 0\right), (\pi, 3),$

$\left(\frac{5\pi}{4}, 0\right), \left(\frac{3\pi}{2}, -3\right), \left(\frac{7\pi}{4}, 0\right), (2\pi, 3)$

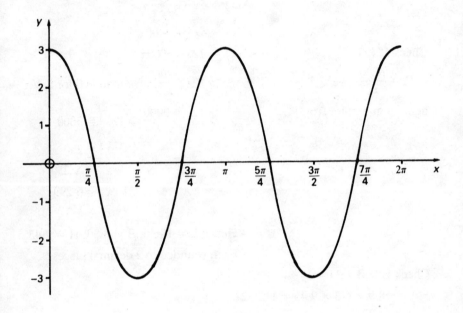

24. *(a)* $f(x) = ax^2 + 4x - 2$ discriminant $b^2 - 4ac$

for equal roots $b^2 - 4ac = 0$

$a = a, b = 4, c = -2$

$b^2 - 4ac$

$= 16 - 4(a)(-2)$

$\Rightarrow 16 + 8a = 0$

$8a = -16$

$a = -2$

$\dfrac{-b \pm \sqrt{0}}{2a}$ $a = -2$

$= \dfrac{-4}{-4} = 1$

$\Rightarrow f(x) = -2x^2 + 4x - 2$

y intercept $(0, -2)$, x intercept $(1, 0)$

$-2x^2$, shape ⌒

(b) $f(x)$ has real roots if $b^2 - 4ac > 0$

$\Rightarrow 16 + 8a > 0$

$8a > -16$

$a > -2$

(c) $f(x)$ has no real roots if $b^2 - 4ac < 0$

$\Rightarrow 16 + 8a < 0$

$8a < -16$

$a < -2$

(d)

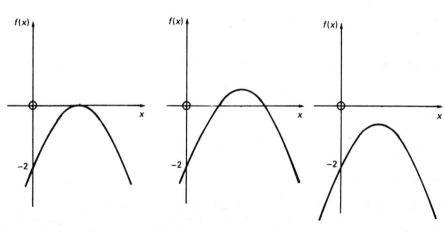

equal roots   real roots   no real roots

25. $3 \cos x - \sin x$ in the form $k \cos (x - \alpha)$

$k = \sqrt{3^2 + (-1)^2}$

$= \sqrt{10}$

$\tan \alpha = \dfrac{-1}{3} \quad \dfrac{\sin -\text{ve}}{\cos +\text{ve}} \text{ in } ④$

$\tan^{-1} \dfrac{1}{3} = 18{\cdot}4$

$\alpha = 360 - 18{\cdot}4 = 341{\cdot}6$

In form $k \cos (x - \alpha)$ gives $\sqrt{10} \cos (x - 341{\cdot}6)$

Zeros of function when $(x - 341{\cdot}6) = 90°$ or $270°$

$\Rightarrow x = 90 + 341{\cdot}6 \qquad\qquad$ or $\qquad\qquad x = 270 + 341{\cdot}6$

$= 431{\cdot}6 - 360 = 71{\cdot}6 \qquad\qquad \Rightarrow 611{\cdot}6 - 360 = 251{\cdot}6$

$x = 0 \qquad \sqrt{10} \cos (0 - 341{\cdot}6) = \sqrt{10} \cos (341{\cdot}6) \doteqdot 3$

maximum $= \sqrt{10}$ when $x = 341{\cdot}6$,

i.e. $\sqrt{10} \cos (341{\cdot}6 - 341{\cdot}6) = \sqrt{10} \cos 0$

minimum $= -\sqrt{10}$ when $x - 341{\cdot}6 = 180 \Rightarrow x = 521{\cdot}6 - 360 = 161{\cdot}6$

Points: $(0, 3), (71{\cdot}6, 0), (161{\cdot}6, -\sqrt{10}), (251{\cdot}6, 0), (341{\cdot}6, \sqrt{10}), (360, 3)$

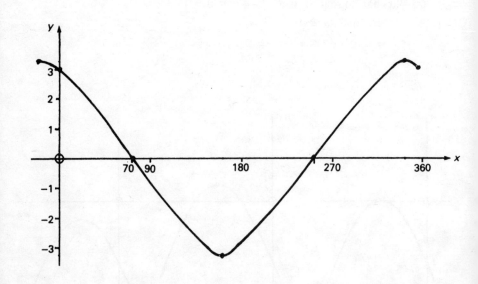